low-dimensional nanoscale systems on discrete spaces

low-dimensional nanoscale systems on discrete spaces

Erhardt Papp
West University of Timisoara, Romania

Codrutza Micu
North University of Baia Mare, Romania

NEW JERSEY • LONDON • SINGAPORE • BEIJING • SHANGHAI • HONG KONG • TAIPEI • CHENNAI

Published by

World Scientific Publishing Co. Pte. Ltd.
5 Toh Tuck Link, Singapore 596224
USA office: 27 Warren Street, Suite 401-402, Hackensack, NJ 07601
UK office: 57 Shelton Street, Covent Garden, London WC2H 9HE

British Library Cataloguing-in-Publication Data
A catalogue record for this book is available from the British Library.

LOW-DIMENSIONAL NANOSCALE SYSTEMS ON DISCRETE SPACES

ISBN-13 978-981-270-638-6
ISBN-10 981-270-638-0

Printed in Singapore by World Scientific Printers (S) Pte Ltd

In the memory of
Siri and Eduard

Preface

Quantum systems on discrete spaces represent to date a rapidly growing research field lying at the interface between tight binding models in solid-state physics and theoretical developments like discrete and q-difference equations. The interplay between such directions provides a deeper understanding of nanoscaled structures which, since more than one decade look increasingly promising for technological applications in electronic devices. Such systems concern the two-dimensional (2D) electron gas under the influence of transversal magnetic fields, the conductor on the 1D lattice in the presence of electric fields, 1D rings in external fields and at last, but not at least, junctions between rings and leads, rings and quantum dots or dots and leads. Promising developments have also been done in the field of quantum LC-circuits. The space discretization has been successful in providing explanations needed for several phenomena like the dynamical localization on the 1D lattice under the influence of time dependent electric fields. It should be remarked, quite satisfactorily, that such phenomena have received evidence in high precision experiments with GaAl/GaAlAs, or other self assembled heterostructures. We also succeeded to establish period doubling in the flux dependent oscillations of total persistent currents in discretized Aharonov-Bohm rings, which deserve further attention.

From the theoretical point of view our main emphasis in this volume is on solvable quantum mechanical systems on the discrete space. Besides the discrete Schrödinger-equation, the most typical example is the second order discrete Harper-equation serving to the description of Bloch-electrons on 2D lattices threaded by a transversal and homogeneous magnetic field. Handling such systems is intimately connected with a lot of mathematical subtleties like orthogonal polynomials of a discrete variable, commensuration and incommensurability effects, periodic and quasiperiodic structures, q-deformed and discrete equations or relationships between localization effects and Lyapunov exponents. Such issues are able to be used for an updated study of thermodynamic and transport properties, but useful details concerning the quantum Hall effect are also included. We would like to emphasize that the presented matter reveals the increasing perspective concerning the role of space-discreteness for a deeper understanding of nanoscaled systems. To this aim mathematical details needed have been introduced in as transparent a manner as possible.

We found it reasonable to restrict ourselves to 11 chapters. In addition, there are more than 300 alphabetically ordered citations, titles of papers included. It is understood that other special matters, such as discrete Klein-Gordon and Dirac equations, or nonlinear lattices, go beyond the immediate scope of this volume. However, the main point remains, namely to solve discrete Schrödinger-equations and especially tight binding models under the influence of external fields. Accordingly, there are reasons to say that we succeeded in reviewing interesting developments within a reasonable amount of space. For the sake of complementary information in the field of semiconductor supperlattices monographs by Bastard (1992), Datta (1995) and Akkermans and Montambaux (2004) are quite useful.

The present volume can be recommended first of all to graduate students and doctoral fellows dealing with Theoretical Physics and its applications. The presented matter is also profitable for advanced students in solid-state physics, applied mathematics and microelectronics. Moreover, useful information for researchers interested in the field is also provided. This volume originates from the updated and revised formulation of our previous lecture in the field (Papp and Micu (2005)).

Interesting discussions with I. Cotaescu, H. Scutaru, I. Bica, D. Grecu, M. Visinescu, E. Burzo, G. Adam, N. Avram, G. Zet, M. Coldea, I. Gottlieb, D. Vangheli, V. Viman and D. Vulcanov, are deeply acknowledged. The technical help by D. Racolta and L. Aur in processing the file is much appreciated. We are also indebted to W. Bestgen, H. D. Doebner, R. M. Weiner and F. Gebhardt for kind hospitality, as well as to CNCSIS/Bucharest for financial support.

Timisoara, Baia Mare *E. Papp*
December 2006 *C. Micu*

Contents

Preface vii

1. **Lattice Structures and Discretizations** 1

1.1 **Discrete derivatives** 1
1.2 **The Jackson derivative** 3
1.3 **The q-integral** 6
1.4 **Generalized q-hypergeometric functions** 7
1.5 **The discrete space-time: a short retrospect** 9
1.6 **Quick inspection of q-deformed Schrödinger equations** 13
1.7 **Orthogonal polynomials of hypergeometric type on the discrete space** 14

2. **Periodic Quasiperiodic and Confinement Potentials** 17

2.1 **Short derivation of the Bloch-theorem** 17
2.2 **The derivation of energy-band structures** 19
2.3 **Direct and reciprocal lattices** 22
2.4 **Quasiperiodic potentials** 25
2.5 **A shorthand presentation of the elliptic Lamè-equation** 27
2.6 **Quantum dot potentials** 28
2.7 **Quantum ring potentials** 31
2.8 **Persistent currents and magnetizations** 32
2.9 **The derivation of the total persistent current for electrons on the 1D ring at $T = 0$** 35
2.10 **Circular currents** 37

3. Time Discretization Schemes 41

3.1 Discretized time evolutions of coordinate and momentum observables . . . 42
3.2 Time independent Hamiltonians of hyperbolic type 43
3.3 Time independent Hamiltonians of elliptic type . . 45
3.4 The derivation of matrix elements . . . 46
3.5 Finite difference Liouville-von Neumann equations and "elementary" time scales . . . 48
3.6 The q-exponential function approach to the q-deformation of time evolution . . . 50
3.7 Alternative realizations of discrete time evolutions and stationary solutions . . . 55

4. Discrete Schrödinger Equations. Typical Examples 57

4.1 The isotropic harmonic oscillator on the lattice . . 58
4.2 Hopping particle in a linear potential . . . 61
4.3 The Coulomb potential on the Bethe-lattice 65
4.4 The discrete s-wave description of the Coulomb-problem . . . 66
4.5 The Maryland class of potentials . . . 69
4.6 The relativistic quasipotential approach to the Coulomb-problem . . . 73
4.7 The infinite square well . . . 75
4.8 Other discrete systems . . . 76

5. Discrete Analogs and Lie-Algebraic Discretizations. Realizations of Heisenberg-Weyl Algebras 79

5.1 Lie algebraic approach to the discretization of differential equations . . . 80
5.2 Describing exactly and quasi-exactly solvable systems . . . 82
5.3 The discrete analog of the harmonic oscillator . . . 84
5.4 Applying the factorization method . . . 87
5.5 The discrete analog of the radial Coulomb-problem 89
5.6 The discrete analog of the isotropic harmonic oscillator . . . 93
5.7 Realizations of Heisenberg-Weyl commutation relations . . . 95

6. Hopping Hamiltonians. Electrons in Electric Field 99

6.1 Periodic and fixed boundary conditions 101
6.2 Density of states and Lyapunov exponents 103
6.3 The localization length: an illustrative example . . 105
6.4 Delocalization effects 107
6.5 The influence of a time dependent electric field . . 108
6.6 Discretized time and dynamic localization 111
6.7 Extrapolations towards more general modulations 114
6.8 The derivation of the exact wavefunction revisited 116
6.9 Time discretization approach to the minimum of the MSD 118
6.10 Other methods to the derivation of the DLC 120
6.11 Rectangular wave fields and other generalizations 122
6.12 Wannier-Stark ladders 125
6.13 Quasi-energy approach to DLC's 126
6.14 The quasi-energy description of dc-ac fields 129
6.15 Establishing currents in terms of the Boltzmann equation 131

7. Tight Binding Descriptions in the Presence of the Magnetic Field 133

7.1 The influence of the nearest and next nearest neighbors 134
7.2 Transition to the wavevector representation 136
7.3 The secular equation 138
7.4 The $Q = 2$ integral quantum Hall effect 140
7.5 Duality properties 142
7.6 Tight binding descriptions with inter-band couplings 143
7.7 Concrete single-band equations and classical realizations 147

8. The Harper-Equation and Electrons on the $1D$ Ring 151

8.1 The usual derivation of the Harper-equation 152
8.2 The transfer matrix 153
8.3 The derivation of Δ-dependent energy polynomials 155
8.4 Deriving Δ-dependent DOS-evaluations 157
8.5 Numerical DOS-studies 160

8.6 **Thermodynamic and transport properties** 161
8.7 **The 1D ring threaded by a time dependent magnetic flux** 167
8.8 **The tight binding description of electrons on the 1D ring** 170
8.9 **The persistent current for the electrons on the 1D discretized ring at $T = 0$** 172

9. **The q-Symmetrized Harper Equation** 175

9.1 **The derivation of the generalized qSHE** 175
9.2 **The three term recurrence relation** 178
9.3 **Symmetry properties** 181
9.4 **The $SL_q(2)$-symmetry of the q SHE** 184
9.5 **Magnetic translations** 188
9.6 **The $SU_q(2)$-symmetry of the usual Harper Hamiltonian** 190
9.7 **Commutation relations concerning magnetic translation operators and the Hamiltonian** 192

10. **Quantum Oscillations and Interference Effects in Nanodevices** 195

10.1 **The derivation of generalized formulae to the total persistent current in terms of Fourier-series** 196
10.2 **The discretized Aharonov-Bohm ring with attached leads** 199
10.3 **Quantum wire attached to a chain of quantum dots** 207
10.4 **Quantum oscillations in multichain nanorings** . . . 210
10.5 **Quantum LC-circuits with a time-dependent external source** 215
10.6 **Dynamic localization effects in L-ring circuits** . . . 219
10.7 **Double quantum dot systems attached to leads** . . 220

11. **Conclusions** 225

11.1 **Further perspectives** 228

Appendix A Dealing with polynomials of a discrete variable 231

Appendix B The functional Bethe-ansatz solution 237

Bibliography 241

Index 259

Chapter 1

Lattice Structures and Discretizations

Quantum systems on discrete spaces have received much interest in many areas like electrons on lattices under the influence of eternal fields, quantum optics, electronics, neuronal networks, signal processing, mesoscopic systems and superlattices. Starting with the discrete space or time one gets faced with the formulation of appropriate discretization schemes. For this purpose discrete variables $x \in \mathbf{Z}$ are almost useful. Other related equations, such as q-difference ones are able to be implied by modern mathematical developments such as quantum groups, the covariant calculus on noncommutative spaces, and R-matrix descriptions (see Chari and Pressley (1994), Faddeev et al (1990)). Here $0 < q < 1$, stands for the generic deformation parameter, but complex q-values and especially roots of unity can also be considered. In addition, there are q-difference equations which are incorporated naturally into to the description of actual physical systems such as Bloch-electrons on two- or three-dimensional lattices threaded by a magnetic field, as already referred to above. Moreover, many systems are able to be characterized by inherent length scales which can be used, at least in principle, for subsequent discretizations, too.

1.1 Discrete derivatives

Assuming a function on a lattice with the spacing a, say $f(x)$, we may introduce left and right hand difference quotients as

$$\delta_- f(x) = \frac{f(x) - f(x-a)}{a} = \frac{1}{a}\left(1 - \exp\left(-a\frac{d}{dx}\right)\right) f(x) \tag{1.1}$$

and

$$\delta_+ f(x) = \frac{f(x+a) - f(x)}{a} = \frac{1}{a}\left(\exp\left(a\frac{d}{dx}\right) - 1\right) f(x) \tag{1.2}$$

respectively. Such quotients, which can also be viewed as discrete derivatives, reproduce the usual one to $O(a^2)$-order. The discretization proceeds in terms of the identification $x = n_i a$, where x_i is an integer, i.e. $n_{i+1} = n_i + 1$. Then the above quotients are replaced by the discrete derivatives

$$\nabla f(x) = f(x) - f(x-1) \tag{1.3}$$

and

$$\Delta f(x) = f(x+1) - f(x) \tag{1.4}$$

respectively, where x stands hereafter for n_i. Accordingly

$$\nabla\Delta f(x) = \Delta\nabla f(x) = f(x+1) - 2f(x) + f(x-1) \tag{1.5}$$

represents the discretized version of the second derivative. On the other hand one starts from the common assumption that the usual dimensionless momentum operator $i\partial/\partial x$ is Hermitian, so that the discrete derivative Δ behaves as $\Delta^+ = -\nabla$ under Hermitian conjugation. Accordingly, the second order discrete derivative operator $\nabla\Delta$ displayed above is itself Hermitian.

We have to remark that the Leibniz-rule for the differentiation of a product is modified as follows

$$\nabla\left(f(x)g(x)\right) = f(x)\nabla g(x) + g(x-1)\nabla f(x) \tag{1.6}$$

and

$$\Delta\left(f(x)g(x)\right) = f(x)\Delta g(x) + g(x+1)\Delta f(x) \tag{1.7}$$

respectively, whereas

$$\Delta^n f(x) = \sum_{k=0}^{n} \frac{n!\,(-1)^k}{(n-k)!k!} f(x+n-k) \tag{1.8}$$

and

$$\nabla^n f(x) = \sum_{k=0}^{n} \frac{n!\,(-1)^k}{k!\,(n-k)!} f(x-k) \quad . \tag{1.9}$$

Such formulae are quite useful for concrete computations. In this context hypergeometric type second-order difference equations like

$$\sigma(x)\nabla\Delta P_n(x) + \tau(x)\Delta P_n(x) + \lambda_n P_n(x) = 0 \tag{1.10}$$

where $\sigma(x)$ ($\tau(x)$) is an at most second-degree (first degree) polynomial, have been studied in some more detail by Nikiforov, Suslov and Uvarov (1991).

It is understood that (1.10) can be rewritten equivalently as a second-order discrete equation:

$$(\sigma(x) + \tau(x))\, P_n(x+1) + \sigma(x) P_n(x-1) + (\lambda - \tau(x) - 2\sigma(x))\, P_n(x) = 0 \tag{1.11}$$

which produces hypergeometric type solutions like Hahn, Chebyshev, Meixner, Krawtchouk or Charlier polynomials for selected forms of $\sigma(x)$ - and $\tau(x)$ -functions (Nikiforov, Suslov and Uvarov (1991)). Main formulae concerning such polynomials are presented in section 1.7 as well as in Appendix A.

1.2 The Jackson derivative

Proceeding in a different manner, Jackson (1909) proposed long ago the generalized q-derivative

$$\partial_x^{(q)} f(x) \equiv \frac{d_q(x)}{d_q x} = \frac{f(qx) - f(x)}{x(q-1)} \tag{1.12}$$

working irrespective of x, which reproduces the usual one if $q \to 1$. Now the Leibniz-rule is modified as

$$\partial_x^{(q)} (f(x)g(x)) = f(x)\partial_x^{(q)} g(x) + g(qx)\partial_x^{(q)} f(x) \tag{1.13}$$

which looks like (1.6). In particular the q-derivative of a power-like function x^n is given by

$$\partial_x^{(q)} x^n = [[n]]_q x^{n-1} \tag{1.14}$$

where

$$[[n]]_q = \frac{q^n - 1}{q - 1} \tag{1.15}$$

denotes the related quantum number. It is clear that $[[n]]_q \to n$ as $q \to 1$. Relatedly, one says that under q-deformation the c-number n is replaced by the quantum-number $[[n]]_q$. We have to be aware that (1.12) opens the way to the formulation of q-analysis. Accordingly special functions are substituted by q ones, and the same concerns, of course the hypergeometric series (Gasper and Rahman (1990)).One realizes that the q-derivative written down above is not unique. Indeed, we can also introduce the symmetrized derivative (see e.g. Kulish and Damaskinsky (1990))

$$\mathcal{D}_x^{(q)} f(x) = \frac{f(qx) - f(q^{-1}x)}{x(q - q^{-1})} \tag{1.16}$$

which yields

$$\mathcal{D}_x^{(q)} x^n = [n]_q x^{n-1} \tag{1.17}$$

instead of (1.14), where now

$$[n]_q = \frac{q^n - q^{-n}}{q - q^{-1}} \quad . \tag{1.18}$$

However, one has

$$[n]_q = q^{1-n} [[n]]_{q^2} \tag{1.19}$$

which shows that such quantum numbers are actually inter-related. Moreover, we have to remark that

$$\Delta f(x) = z(q-1)\, \partial_z^{(q)} \varphi(z) \tag{1.20}$$

where $z = q^x$ and $f(x) = \varphi(z)$. This relationship is able to serve to the conversion of discrete equations into a q-difference ones and conversely.

Needless to say that q-analysis is far from being trivial. So, the q-binomial theorem reads (see Bailey (1935))

$$\sum_{n=0}^{\infty} \frac{(a;q)_n}{(q;q)_n} x^n = \frac{(ax;q)_\infty}{(x;q)_\infty} \tag{1.21}$$

in which "a" denotes a constant parameter. Accordingly, a q-analog of the gamma function is given by

$$\Gamma_q(x) = \frac{(q;q)_\infty}{(q^x;q)_\infty} (1-q)^{1-x} \tag{1.22}$$

which has also been expressed before in terms of the q-integral proposed by Jackson (1910). For the convergence sakes one should consider that $0 < q < 1$ both in (1.21) and (1.22). In the above formulae $(a;q)_n$ denotes the q-shifted factorial

$$(a;q)_n = (1-a)(1-aq)\cdots\left(1-aq^{n-1}\right) \tag{1.23}$$

where $(a;q)_0 = 1$ so that

$$(a;q)_\infty = \prod_{n=0}^{\infty} (1-aq^n) \quad . \tag{1.24}$$

We have to remark that the present q-Gamma function satisfies the typical functional equation

$$\Gamma_q(x+1) = [[x]]_q \Gamma_q(x) \tag{1.25}$$

which confirms in turn the relevance of its definition via (1.22). The infinite products quoted above yield a shifted factorial as follows

$$(a;q)_n = \frac{(a;q)_\infty}{(aq^n;q)_\infty} \tag{1.26}$$

which is now valid for arbitrary complex n-values.

Before proceeding further we would like to say that the Jackson derivative has also been rediscovered recently in terms of the radial reduction of the covariant derivative characterizing the quantum-group $SO_q(N)$, as illustrated by q-deformed radial Schrödinger equations established before (Carow-Watamura and S. Watamura (1997), Papp (1997)).

1.3 The q-integral

In order to perform scalar products and suitable normalizations we have to apply the Jackson q-integral (Gasper and Rahman (1990), Jackson (1910)). So one has

$$\int_0^a f(x)\, d_q x = a(1-q) \sum_{k=0}^{\infty} f\left(aq^k\right) q^k \tag{1.27}$$

if $0 < q < 1$, but

$$\int_0^a f(x)\, d_q x = a(1-q) \sum_{k=0}^{\infty} \frac{1}{q^{k+1}} f\left(\frac{a}{q^{k+1}}\right) \tag{1.28}$$

if $q > 1$. Inserting e.g. $f(x) = x^n$ yields

$$\int_0^a x^n\, d_q x = \frac{a^{n+1}}{[[n+1]]_q} \tag{1.29}$$

in both cases. One sees that (1.14) and (1.29) express inverse operations, as one might expect. Furthermore, there is

$$\int_0^{\infty} f(x)\, d_q x = (1-q) \sum_{k=-\infty}^{+\infty} f\left(q^k\right) q^k \tag{1.30}$$

if $0 < q < 1$, which can be generalized on the whole real axis as

$$\int_{-\infty}^{+\infty} f(x)\, d_q x = (1-q) \sum_{k=-\infty}^{+\infty} \left[f\left(q^k\right) + f\left(-q^k\right)\right] q^k \quad . \tag{1.31}$$

Of course, one realizes that convergence conditions $0 < q < 1$ and $q > 1$ mentioned above can be generalized as $|q| < 1$ and $|q| > 1$, respectively.

The symmetrized versions of (1.27) and (1.28) are given by

$$\int_0^a f(z)\, D_q z = \left(\frac{1}{q} - q\right) \sum_{k=0}^{\infty} aq^{2k+1} f\left(aq^{2k+1}\right) \tag{1.32}$$

and

$$\int_0^a f(z)\, D_q z = \left(q - \frac{1}{q}\right) \sum_{k=0}^{\infty} \frac{a}{q^{2k+1}} f\left(\frac{a}{q^{2k+1}}\right) \tag{1.33}$$

if $|q| < 1$ and $|q| > 1$, respectively. So, one obtains

$$\int_0^a z^n D_q z = \frac{a^{n+1}}{[n+1]_q} \tag{1.34}$$

which is valid again in both cases. This reproduces, of course,(1.29) up to the selection of the appropriate quantum number. Integrations by parts can also be easily done. Using (1.13) one finds e.g.

$$\int_0^a f d_q g = f(x)\, g(x)|_0^a - \int_0^a g(qx)\, d_q f \tag{1.35}$$

which works both for $|q| < 1$ and $|q| > 1$. One remarks that the discrete counterpart of (1.35) reads

$$\sum_{x=a}^{b-1} f(x)\, \Delta g(x) = f(x)\, g(x)|_a^b - \sum_{x=a}^{b-1} g(x+1)\, \Delta f(x) \tag{1.36}$$

where x stands again for the discrete variable and which proceeds in accord with (1.7). A similar equation can be derived for the left difference operator ∇.

1.4 Generalized q-hypergeometric functions

Many solutions of discrete equations are expressed in terms of the generalized hypergeometric function (Gasper and Rahman (1990), Gradshteyn and Ryzhik (1965))

$${}_{p_1}F_{p_2}\left(\begin{array}{c} \alpha_1\,,\, \alpha_2\,,\, \ldots\,,\, \alpha_{p_1} \\ \\ \beta_1\,,\, \beta_2\,,\, \ldots\,\, \beta_{p_2} \end{array} \middle|\; z \right) = \tag{1.37}$$

$$= \sum_{k=0}^{\infty} \frac{(\alpha_1)_k \cdots (\alpha_{p_1})_k}{(\beta_1)_k \cdots (\beta_{p_1})_k} \frac{z^k}{k!}$$

where p_1 and p_2 are indices and where α_j and β_j are parameters. The classical shifted factorial is given by

$$(a)_k = \frac{\Gamma(a+k)}{\Gamma(a)} = a(a+1)\cdots(a+k-1) \tag{1.38}$$

as usual, where $(a)_0 = 1$. Recall that the usual Gaussian hypergeometric function is

$$F(\alpha,\beta,\gamma;z) = {}_2F_1\left(\begin{matrix}\alpha\ ,\ \beta\\ \\ \gamma\end{matrix}\middle|\ z\right) \tag{1.39}$$

whereas the confluent one is given by

$$F(\alpha,\gamma;z) = {}_1F_1\left(\begin{matrix}\alpha\\ \\ \gamma\end{matrix}\middle|\ z\right) \quad . \tag{1.40}$$

Resorting for convenience, to the quantum number (1.15), we shall perform the q-generalization of (1.37) in terms of the substitutions

$$(a)_k \to \frac{(q^a,q)_k}{(1-q)^k} = \prod_{l=0}^{k-1}[[a+l]]_q \tag{1.41}$$

and

$$k! \to [[k]]_q! = \frac{(q,q)^k}{(1-q)^k} = \prod_{l=1}^{k}[[l]]_q \quad . \tag{1.42}$$

This leads to the q-generalization

$${}_{p_1}F_{p_2}^{(q)}\left(\begin{matrix}\alpha_1\ ,\ \alpha_2\ ,\ \ldots\ ,\ \alpha_{p_1}\\ \\ \beta_1\ ,\ \beta_2\ ,\ \ldots\ \ \beta_{p_2}\end{matrix}\middle|\ \ \widetilde{z}\ \right) = \tag{1.43}$$

$$= \sum_{k=0}^{\infty}\frac{(q^{\alpha_1},q)_k\cdots(q^{\alpha_{p_1}},q)_k}{(q^{\beta_1},q)_k\cdots(q^{\beta_{p_2}},q)_k}\frac{\widetilde{z}^k}{(q,q)_k}$$

which reproduces (1.37) as $q \to 1$, where

$$\widetilde{z} = (1-q)^{p_2-p_1+1}z \quad . \tag{1.44}$$

In order to identify the generalized q-hypergeometric function it is convenient to rewrite (1.43) as follows

$$ {}_{p_1}F_{p_2}^{(q)}\left(\begin{matrix}\alpha_1\ ,\alpha_2\ ,\ldots\ ,\ \alpha_{p_1}\\ \\ \beta_1\ ,\beta_2\ ,\ldots\ \ \beta_{p_2}\end{matrix}\ \middle|\ \gamma\,\tilde{z}\right) = \tag{1.45} $$

$$ = \sum_{k=0}^{\infty} C_k \tilde{z}^k $$

where γ is an additional factor. One would then obtain the recurrence relation

$$ C_{k+1} = \gamma \frac{(1-q)^{p_2-p_1+1}\left(1-q^{\alpha_1+k}\right)\cdots\left(1-q^{\alpha_{p_1}+k}\right)}{\left(1-q^{k+1}\right)\left(1-q^{\beta_1+k}\right)\cdots\left(1-q^{\beta_{p_2}+k}\right)} C_k \tag{1.46} $$

which can also be used in order to check the appearance of polynomial solutions for which $C_{n+1} = 0$. The q-exponential function is also of a special interest, but it will be discussed in section 3.6 in a close connection with the q-deformation of the time evolution.

1.5 The discrete space-time: a short retrospect

The basic idea about a discrete space and/or time is as old as the very beginning of the scientific thinking. The old Greeks speculated about the generalization of the atomistic structure of matter to space-time, as illustrated in an excellent volume by Vialtzew (1965). Such steps deserve an actual appreciation even from the perspective of contemporary physics. With the advent of quantum mechanics, the question of whether subdivisions of space and time intervals can be indefinitely performed or not, has focussed increasing interest. The latter alternative, which leads to the introduction of space-time quanta proceeding in a close connection with ultimate accuracies of space-time measurements, is looks almost promising. Pioneering work along this direction has been done by many authors, but here we would like just to remember specifically contributions done by Poincaré (1913), Planck (1913), Ambarzumian and Iwanenko (1930), March (1937), Heisenberg (1938a,1938b), Wheeler (1957), Brill and Gowdy (1970) and Finkelstein (1997). In "Physics and Reality" Einstein said:

"To be sure, it has been pointed out that the introduction of a spacetime continuum may be considered as contrary to nature in view of the molecular

structure of everything which happens on a small scale. It is maintained that perhaps the success of the Heisenberg method points to a purely algebraical method of description of nature, that is to the elimination of continuum functions from physics. Then, however, we must also give up, by principle, the spacetime continuum" (see also Atakishiyev and Suslov (1990)).

Further developments rely on a celebrated paper by Snyder (1947) dealing with the introduction of a Lorentz invariant space-time. We then have to consider a discrete space-time in which there exists a smallest unit of length, say $a \neq 0$. The basic expectation is that accounting for a such length leads to the elimination of divergence difficulties plaguing quantum field theory. Introducing a such length in space-time yields, however, non-commutative relationships between x, y, z and t, so that space-time coordinates have to be considered by now as observables. More exactly, they are identified with the generators of the group of transformations leaving invariant the five-dimensional de Sitter quadratic form (see also Yang (1947))

$$R_5^2 = -\eta_0^2 + \eta_1^2 + \eta_2^2 + \eta_3^2 + \eta_4^2 \tag{1.47}$$

in which the η's are assumed to be real variables. One would then obtain Hermitian space-time generators as follows

$$X_1 = ia\left(\eta_4 \frac{\partial}{\partial \eta_1} - \eta_1 \frac{\partial}{\partial \eta_4}\right) \tag{1.48}$$

$$X_2 = ia\left(\eta_4 \frac{\partial}{\partial \eta_2} - \eta_2 \frac{\partial}{\partial \eta_4}\right) \tag{1.49}$$

$$X_3 = ia\left(\eta_4 \frac{\partial}{\partial \eta_3} - \eta_3 \frac{\partial}{\partial \eta_4}\right) \tag{1.50}$$

and

$$\widetilde{T}_0 = \frac{ia}{c}\left(\eta_4 \frac{\partial}{\partial \eta_0} - \eta_0 \frac{\partial}{\partial \eta_4}\right) \quad . \tag{1.51}$$

In addition, there are three operators like

$$L_k = -i\hbar\varepsilon_{kls}\eta_l \frac{\partial}{\partial \eta_s} \tag{1.52}$$

where $k, l, s = 1, 2, 3$ and where the usual summation convention is assumed. These latter operators correspond to the . Further three operators like

$$M_1 = i\hbar \left(\eta_0 \frac{\partial}{\partial \eta_1} + \eta_1 \frac{\partial}{\partial \eta_0} \right) \tag{1.53}$$

$$M_2 = i\hbar \left(\eta_0 \frac{\partial}{\partial \eta_2} + \eta_2 \frac{\partial}{\partial \eta_0} \right) \tag{1.54}$$

and

$$M_3 = i\hbar \left(\eta_0 \frac{\partial}{\partial \eta_3} + \eta_3 \frac{\partial}{\partial \eta_0} \right) \tag{1.55}$$

bearing on the Lorentz-boosts, have also to be considered. Accordingly, one finds the commutation relations

$$[X_k, X_l] = i\frac{a^2}{\hbar}\varepsilon_{kls}L_s \tag{1.56}$$

and

$$\left[\widetilde{T}_0, X_k\right] = i\frac{a^2}{\hbar c}M_k \tag{1.57}$$

which are responsible for the discrete space-time and which reproduce the space-time continuum as soon as $a \to 0$. It is understood that L_k and M_k commutes with the Minkowski quadratic form

$$S_4^2 = c^2\widetilde{T}_0^2 - X_1^2 - X_2^2 - X_3^2 \tag{1.58}$$

provided (1.48)-(1.51) are valid. This means that the discrete space-time defined in this manner is itself Lorentz-invariant.

The eigenvalues, say x'_k, characterizing the discrete space are given by

$$x'_k = n_k a \tag{1.59}$$

where $n_k \in \mathbb{Z}$, which reveals the space discreteness in an explicit manner. The $\widetilde{T}_0$ operator has, however, a continuous spectrum, which shows, so far, that the discreteness of space does not imply necessarily a discrete time. In this context a continuous time can be preserved, however, for reasons of mathematical simplicity. Either case, (1.59) shows that there are actual interplays between relativity and space-discreteness. Several ideas and approaches to the discrete space have also been sketched in an unpublished paper by Heisenberg in 1930 (see also Carazza and Kragh (1995)). In this context the fundamental length has also been conceived as the absolute minimum of the position uncertainty of a particle.

Other manifestations of the fundamental length have been done by defining covariant space-time and four-momentum operators as (Hellund and Tanaka (1954))

$$X_\mu = \widetilde{x}_\mu - a^2 \frac{\partial}{\partial \widetilde{x}_\mu} \widetilde{x}_\nu \frac{\partial}{\partial \widetilde{x}_\nu} \tag{1.60}$$

and

$$P_\mu = \widetilde{p}_\mu - i\hbar \frac{\partial}{\partial \widetilde{x}_\mu} \tag{1.61}$$

respectively, where $\widetilde{x}_\mu$ and $\widetilde{p}_\mu$ are c-numbers like $\widetilde{x}_\mu = (x, y, z, ict)$ and $\widetilde{p}_\mu = (p_x, p_y, p_z, iE/c)$. For convenience, the same quotations for space-time operators have been preserved. This yields the commutation relations

$$[X_\mu, P_\nu] = i\hbar \left(\delta_{\mu\nu} + \frac{a^2}{\hbar^2} P_\mu P_\nu \right) \tag{1.62}$$

in which case Heisenberg's uncertainty principle is generalized as follows

$$\Delta X_\mu \Delta P_\nu \geq \hbar \left(1 + \frac{a^2}{\hbar^2} P_\mu^2 \right) \tag{1.63}$$

for $\mu = \nu$ (no summation). Moreover, there is

$$\Delta X_1 \Delta \widetilde{T}_0 \geq \frac{a^2}{c} \tag{1.64}$$

which shows that elementary space-time volumes have also to be considered.

Other evolutions could also be mentioned (see Kadyshevsky (1961)), but the main point is that accounting for (1.59) leads to the introduction of discrete-equations serving as starting points for an improved theoretical descriptions in several respects. Interestingly enough, similar results are provided by q-deformed oscillators for which q is a root of unity, as shown by Bonatsos et al (1994). Then position and momentum operators have discrete eigenvalues, which are the roots of certain q-deformed Hermite polynomials. This leads, in general, to a non-equidistant phase-space lattice. An exception is the parafermionic oscillator, for which an equidistant space-phase lattice can be readily derived.

In other words, there are theoretical supports to the existence of a discrete space in which the fundamental length plays a role which is similar, in a way or another, to the lattice spacing in solid state physics. Accordingly, discrete derivatives written down in section 1.1 are well motivated from a quantum mechanical point of view. From the mathematical perspective, similar developments can be traced back to the introduction of q-hypergeometric functions and Jackson-derivatives (see e.g. chapter 1 in Gasper and Rahman (1990)).

1.6 Quick inspection of q-deformed Schrödinger equations

The q-deformed Schrödinger-equation on the non-commutative quantum Euclidian space has been analyzed for the harmonic oscillator (see e.g. Carow-Watamura and Watamura (1994)), for the Coulomb potential (Song and Liao (1992)), Feigenbaum and Freund (1996), Micu (1999)), as well as for the free particle (Hebecker and Weich (1992), Ocampo (1996), Papp (1997)). For this purpose one resorts to the q-deformed Laplacian, which has been derived in terms of the $SO_q(N)$-covariant calculus (Carow-Watamura and S. Watamura (1994)). Just mention that the non-commutative N-dimensional quantum Euclidian space is characterized by typical coordinate dependent realizations of $[x_j, x_k]$ commutators in terms of bilinear or quadratic monomials such as given e.g. by the $SO_q(3)$ relationships $[x_1, x_2] = (q-1)x_2x_1$, $[x_2, x_3] = (q-1)x_3x_2$ and $[x_1, x_3] = (1-q)x_2^2/\sqrt{q}$. In addition, there are well defined q-difference realizations of annihilation and creation operators, such as done in the study of a different version of the q-deformed harmonic oscillator (Atakishiyev and Suslov (1990), Li and Sheng (1992)). An interesting q-deformation of the Schrödinger-equation has been established by resorting to a q-deformation of the Witt algebra (Twarock (1999)), in which case the nonlinear Schrödinger-equation derived by Doebner and Goldin (1992) gets reproduced as $q \to 1$. Other issues, such as q-deformations of the phase-space (Zumino (1991), Ubriaco (1993), Kehagias and Zoupanos (1994), Celeghini et al (1995), Wess (1997)), can also be mentioned. We have to recognize, however, that a sound confirmation of the physical relevance of q-deformations referred to above is still lacking. Nevertheless, there are parameter dependent difference equations in which the q-parameter is implemented from the very beginning in terms of an appropriate physical description, such as the q-symmetrized Harper-equation (see chapter 9).

1.7 Orthogonal polynomials of hypergeometric type on the discrete space

Classical orthogonal polynomials of a discrete variable are of much interest for several applications. A systematic study of such polynomials has already be done by Nikiforov, Suslov and Uvarov (1991). These polynomials are specified by the $\sigma(x)$- and $\tau(x)$-functions in (1.10). In the case of hypergeometric type polynomials we have to choose $\sigma(x)$ and $\tau(x)$ as polynomials of at most second and of first degrees in x, respectively. So one has e.g.

$$\sigma(x) = x, \qquad \tau(x) = \mu(\gamma - x) - x, \qquad x \in [0, b] \tag{1.65}$$

for Krawtchouk polynomials, whereas

$$\sigma(x) = x, \qquad \tau(x) = \mu(\gamma + x) - x, \qquad x \in [0, \infty) \tag{1.66}$$

for Meixner-ones. In the first (second) case one has $\gamma = N_0 = b - 1$, $\mu = p/q$ with $p + q = 1$ ($0 < \mu < 1$ and $\gamma > 0$). Equation (1.10) can also be converted into the self-adjoint form

$$\Delta\left(\sigma(x)\rho(x)\nabla y(x)\right) + \lambda\rho(x)y(x) = 0 \tag{1.67}$$

where λ denotes the eigenvalue and where the weight function $\rho(x)$ is implemented via

$$\Delta(\sigma(x)\rho(x)) = \tau(x)\rho(x) \quad . \tag{1.68}$$

In addition, (1.67) can be generalized as

$$\Delta\left(\sigma(x)\rho_m(x)\nabla v_m(x)\right) + \mu_m\rho_m(x)v_m(x) = 0 \tag{1.69}$$

where

$$v_m(x) = \Delta^m y(x) \tag{1.70}$$

and where m is a positive integer. Accordingly, (1.68) becomes

$$\Delta(\sigma(x)\rho_m(x)) = \tau_m(x)\rho_m(x) \tag{1.71}$$

where

$$\tau_m(x) = \tau(x+m) + \sigma(x+m) - \sigma(x) \quad . \tag{1.72}$$

The generalized weight function reads

$$\rho_m(x) = \sigma(x+1)\rho_{m-1}(x) = \rho(x+m)\prod_{l=1}^{m}\sigma(x+l) \tag{1.73}$$

for which the eigenvalue becomes

$$\mu_m = \mu_{m-1} + \Delta\tau_{m-1}(x) \quad . \tag{1.74}$$

Furthermore, one finds the concrete realizations

$$\tau_m(x) = \tau(x) + m(\sigma\prime(x) + \tau\prime(x)) + \frac{m^2}{2}\sigma\prime\prime(x) \tag{1.75}$$

and

$$\mu_m = \lambda + m\tau\prime(x) + \frac{m}{2}(m-1)\sigma\prime\prime(x) \quad . \tag{1.76}$$

Other details are presented for interested readers in Appendix A.

Chapter 2

Periodic Quasiperiodic and Confinement Potentials

Schrödinger equations with periodic potentials concern both continuous and discrete spaces, but the latter choice seems to be almost suitable from a general theoretical point of view. Indeed periodic lattice structures, such as the ones from solid state physics, give rise to actual periodic potentials, which produce in turn typical manifestations, such as the appearance of energy bands. The main ingredient is the Bloch-theorem (see also Ashcroft and Mermin (1976)) working in conjunction with the selection of the first Brillouin zone. Even if they are not an absolute novelty, such issues are well suited for a safe explanation of energy-bands, of persistent currents or of dynamic localization effects concerning low dimensional systems on discrete spaces. Preliminaries referring to confinement potentials for quantum dots and rings, as well as to persistent currents, are also addressed.

2.1 Short derivation of the Bloch-theorem

Let us consider the one-dimensional Schrödinger equation

$$\mathcal{H}\psi(x) = -\frac{\hbar^2}{2m_0}\frac{d^2\psi}{dx^2} + V(x)\psi(x) = E\psi(x) \tag{2.1}$$

in which there is

$$V(x+a) = V(x) \tag{2.2}$$

irrespective of x. One then says that the potential is periodic, the period being denoted by a. The crystalline structure can be accounted for by inserting the effective mass m^* instead of m_0. Here the coordinate is continuous, but dealing with the discrete space one inserts $x = na$, where n

is an integer while a stands for the lattice spacing. Looking for a square integrable wavefunction for which

$$\psi(x) \in \{L_2(-\infty, +\infty),\ dx\} \tag{2.3}$$

one sees immediately that the same holds for $\psi(x+a)$. Accordingly the translated wavefunction

$$\psi(x+a) \equiv T_0(a)\psi(x) = \exp(if(a))\,\psi(x) \tag{2.4}$$

is the by-product of an unitary transformation, in which case $f(a)$ is a pure phase. For definiteness, this phase may be located on the unit circle such that $f(a) \in [-\pi, \pi]$. The translation operator is given by

$$T_o(a) = \exp(a\partial/\partial x) = \exp(iap_{op}) \tag{2.5}$$

where $p_{op} = -i\partial/\partial x$ $(\hbar = 1)$. Of course, one has the commutation relation

$$[H, T_0(a)] = 0 \tag{2.6}$$

which proceeds in accord with (2.2).

Repeating the translation n-times one finds

$$\psi(x+na) = \exp(i\ nf(a))\,\psi(x) \tag{2.7}$$

but one should also have

$$\psi(x+na) = \exp(i\ f(na))\,\psi(x) \tag{2.8}$$

as the latter translation proceeds in terms of the rescaled period $a\prime = na$. Furthermore, there is $f(na) = nf(a)$, so that nothing prevents us from doing the identification

$$f(a) = ka \in [-\pi, \pi] \quad . \tag{2.9}$$

It is clear that the wavenumber k quoted above has been invoked on dimensionality grounds. So, the normalized eigenfunction of (2.1) exhibits itself a certain kind of twisted periodic behavior like

$$\psi(x+a) \equiv \psi_k(x+a) = \exp(ika)\,\psi_k(x) \quad . \tag{2.10}$$

This reproduces the $1D$ Bloch-theorem (see Ashcroft and Mermin (1976), Flügge (1971)) and corresponds, as it stands, to a forward translation. We have to remark that a modulated plane wave:

$$\psi_k(x) = \exp(ikx)\ u_k(x) \tag{2.11}$$

obeys (2.2) if

$$u_k(x+a) = u_k(x) \tag{2.12}$$

which shows that the modulation should be periodic, too. As a corollary, the energy becomes itself both k-dependent and periodic:

$$E = E(k) = E\left(k + \frac{2\pi}{a}\right) \tag{2.13}$$

where $2\pi/a$ has the meaning of a period characterizing the very existence of energy bands (see (2.24)).

2.2 The derivation of energy-band structures

In order to illustrate the appearance of energy bands, let us consider the periodic potential produced by an infinite sequence of identical barriers like (see also Davison and Steslicka (1992))

$$V(x) = \begin{cases} U_0\ , & x \in [na_1 + (n-1)\,b_1,\ n\,(a_1+b_1)] \\ \\ 0\ , & x \notin [na_1 + (n-1)\,b_1,\ n\,(a_1+b_1)] \end{cases} \tag{2.14}$$

where $n = 0, \pm 1, \pm 2, \ldots$. One sees that

$$V(x + a_1 + b_1) = V(x) \tag{2.15}$$

so that now $a = a_1 + b_1$ plays the role of the lattice spacing. The depth of the potential barrier is b_1, the length of the zero-potential cell is a_1, whereas $U_0 > 0$ is a constant amplitude.

Next, let us choose three adjacent zones, say $\mathcal{D}_1$, $\mathcal{D}_2$, and $\mathcal{D}_3$, for which $x \in [-b_1, 0]$, $x \in [0, a_1]$ and $x \in [a_1, a_1 + b_1]$, respectively. Assuming at the beginning that $E > U_0$, one finds the solutions

$$\psi(x) = \psi_1(x) = A_1 \exp(ik_1 x) + B_1 \exp(-ik_1 x) \tag{2.16}$$

and

$$\psi(x) = \psi_2(x) = A_2 \exp(ik_2 x) + B_2 \exp(-ik_2 x) \tag{2.17}$$

in $\mathcal{D}_1$ and $\mathcal{D}_2$, where

$$k_1 = \sqrt{\frac{2m}{\hbar^2}(E - U_0)} \tag{2.18}$$

and

$$k_2 = \sqrt{\frac{2m}{\hbar^2} E} \tag{2.19}$$

respectively. On the other hand, one has

$$\psi(x) = \psi_3(x) = \exp(ikx)\, u_k(x) \tag{2.20}$$

for $x \in \mathcal{D}_3$, in which

$$u_k(x) = u_k(x - a) = \exp(-ik(x - a))\, \psi_1(x - a) \quad . \tag{2.21}$$

Putting together (2.16) and (2.20), one finds immediately that

$$\psi_3(x) = \exp(ika)\left[A_1 \exp(ik_1(x - a)) + B_1 \exp(-ik_1(x - a))\right] \tag{2.22}$$

for $x \in \mathcal{D}_3$. Next, we have to apply continuity conditions concerning the wavefunction and its first derivative at $x = 0$ and $x = a_1$. This results in four homogeneous matching equations for the amplitudes A_1, B_1, A_2 and B_2:

$$\begin{pmatrix} 1 & 1 & -1 & -1 \\ \exp(i(\theta - \theta_1)) & \exp(i(\theta + \theta_1)) & -\exp(i\theta_2) & -\exp(-i\theta_2) \\ k_1 & -k_1 & -k_2 & k_2 \\ k_1 \exp(i(\theta - \theta_1)) & -k_1 \exp(i(\theta + \theta_1)) & -k_2 \exp(i\theta_2) & k_2 \exp(-i\theta_2) \end{pmatrix} . \tag{2.23}$$

$$\begin{pmatrix} A_1 \\ B_1 \\ A_2 \\ B_2 \end{pmatrix} = 0$$

where $\theta_1 = k_1 b_1$, $\theta_2 = k_2 a_1$ and $\theta = ka$. Denoting by $\mathcal{M}$ the matrix just displayed above, one finds the eigenvalue-condition

$$P\left(E; \theta_1, \theta_2\right) = \cos\theta \tag{2.24}$$

via $\det(\mathcal{M}) = 0$, where

$$P\left(E; \theta_1, \theta_2\right) = \cos\theta_1 \cos\theta_2 - \frac{k_1^2 + k_2^2}{2k_1 k_2} \sin\theta_1 \sin\theta_2 \tag{2.25}$$

and where the r.h.s. of (2.24) agrees with (2.13). The interesting point concerning (2.24) is the remarkable separation of the θ-dependent term, i.e. of $\cos\theta$, on the r.h.s. of the equation. It is obvious that k_1 should be replaced by iK_1, if $0 < E < U_0$, where

$$K_1 = \sqrt{\frac{2m}{\hbar^2}\left(U_0 - E\right)} \quad . \tag{2.26}$$

Then (2.25) becomes

$$P\left(E; i\widetilde{\theta}_1, \theta_2\right) = \cosh\widetilde{\theta}_1 \cos\theta_2 - \frac{k_2^2 - K_1^2}{2K_1 k_2} \sinh\widetilde{\theta}_1 \sin\widetilde{\theta}_{22} \tag{2.27}$$

where $\widetilde{\theta}_1 = K_1 b_1$. Invoking (2.8), one finds that unions over admissible energies are done by the inequalities

$$-1 \leq P\left(E; \theta_1, \theta_2\right) \leq 1 \tag{2.28}$$

and

$$-1 \leq P\left(E; i\widetilde{\theta}_1, \theta_2\right) \leq 1 \tag{2.29}$$

respectively. In other words, we are ready to derive energy band structures for which the band edges are produced by the equality signs in (2.28) and (2.29). Conversely, one finds energy gaps in the complementary regions, i.e. at energies for which $\left|P\left(E; \theta_1, \theta_2\right)\right| > 1$ and $\left|P(E; i\widetilde{\theta}_1, \theta_2)\right| > 1$, respectively.

Using for instance dimensionless variables like $\gamma_1 = \sqrt{2}b_1/l_c$, $\gamma_2 = \sqrt{2}a_1/l_c$, $W = E/mc^2$ and $u_0 = U_0/mc^2$ one sees that the energy-function exhibits the form

$$P(E;\theta_1,\theta_2) = \cos\left(\gamma_1\sqrt{W-u_0}\right)\cos\left(\gamma_2\sqrt{W}\right) - \tag{2.30}$$
$$-\frac{2W-u_0}{\sqrt{(W-u_0)W}}\sin\left(\gamma_1\sqrt{W-u_0}\right)\sin\left(\gamma_2\sqrt{W}\right)$$

where $l_c = \hbar/mc$ denotes the Compton wavelength. In particular, the W-dependence of $P(E;\theta_1,\theta_2)$ is displayed in figure 2.1, for $\gamma_1 = 2$, $\gamma_2 = 3$, $u_0 = 1$ and $0 < W < 10$. We have to remark that (2.28) and (2.29) yield four bands expressed approximately by the union

$$W \in [0.24,\ 0.57] \cup [1.08,\ 1.82] \cup [3.65,\ 4.86] \cup [7.23,\ 10] \tag{2.31}$$

in which the individual bandwidths increase with W.

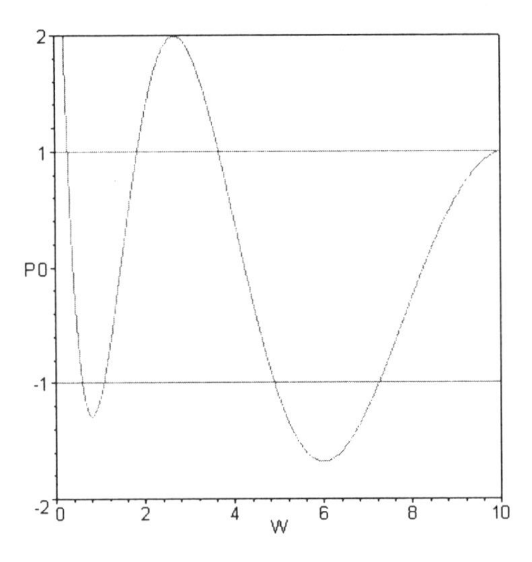

Fig. 2.1 **The W-dependence of $P\left(E;\widetilde{\theta}_1,\widetilde{\theta}_2\right)$ for $\gamma_1 = 1$, $\gamma_2 = 3$ and $u_0 = 1$.**

2.3 Direct and reciprocal lattices

In solid state physics one gets faced with configuration space vectors like:

$$\vec{R}_n = n_1 \vec{a}_1 + n_2 \vec{a}_2 + n_3 \vec{a}_3 \tag{2.32}$$

referred to as Bravais-lattice vectors, where $n_j \in$

Z (j=1,2,3) and where the $\vec{a}_j$'s are responsible for the primitive unit cell. Finite translations in the real space are then described by

$$\mathbf{T}_0 \left(\vec{R}_n \right) = \exp \left(i \vec{p}_{op} \cdot \vec{R}_n \right) \tag{2.33}$$

where $\vec{p}_{op} = -i\nabla$. This yields

$$\mathbf{T}_0 \left(\vec{R}_n \right) \psi_{\vec{k}} \left(\vec{x} \right) \equiv \psi_{\vec{k}} \left(\vec{x} + \vec{R}_n \right) = \exp \left(i \vec{k} \cdot \vec{R}_n \right) \psi_{\vec{k}} \left(\vec{x} \right) \tag{2.34}$$

where $\vec{k} \equiv \vec{k}_B$ is the Brillouin wavevector which incorporates the three-dimensional version of Bloch's theorem (see e.g. Ashcroft and Mermin (1976)).

The reciprocal lattice is characterized by the primitive vectors

$$\vec{g}_j = 2\pi \varepsilon_{jkl} \frac{\vec{a}_k \times \vec{a}_l}{V_1^{(3)}} \tag{2.35}$$

where $V_1^{(3)} = \vec{a}_1 \cdot (\vec{a}_2 \times \vec{a}_3)$ is the volume of the unit cell. A general reciprocal lattice vector is then given by

$$\vec{K}_m = m_1 \vec{g}_1 + m_2 \vec{g}_2 + m_3 \vec{g}_3 \tag{2.36}$$

where m_j' s are again integers. One sees that $\vec{a}_j \cdot \vec{g}_l = \delta_{jl}$, so that

$$\vec{R}_n \cdot \vec{K}_m = 2\pi n' \tag{2.37}$$

where

$$n' = n_1 m_1 + n_2 m_2 + n_3 m_3 \in \mathbb{Z} \quad . \tag{2.38}$$

Defining the wavefunction in the complementary wavevector representation by the discrete Fourier-transform

$$\widetilde{\psi}_F \left(\vec{k'} \right) = \sum_{\vec{R}_n} \psi \left(\vec{R}_n \right) \exp \left(-i \vec{k'} \cdot \vec{R}_n \right) \tag{2.39}$$

one has

$$\widetilde{\psi}_F\left(\overrightarrow{k'}+\overrightarrow{k}_B\right)=\widetilde{\psi}\left(\overrightarrow{k'}\right) \tag{2.40}$$

where

$$\widetilde{\psi}\left(\overrightarrow{k'}\right)=\sum_{\overrightarrow{R}_n}\psi\left(\overrightarrow{R}_n\right)\exp\left(-i\left(\overrightarrow{k'}+\overrightarrow{k}_B\right)\cdot\overrightarrow{R}_n\right) \quad . \tag{2.41}$$

The sum in (2.39) and (2.41) is over all n_j's. This shows the occurrence of a periodic behavior like

$$\widetilde{\psi}\left(\overrightarrow{k'}+\overrightarrow{K}_m\right)=\widetilde{\psi}\left(\overrightarrow{k'}\right) \tag{2.42}$$

in which case

$$\widetilde{\psi}_F\left(\overrightarrow{k'}+\overrightarrow{k}_B+\overrightarrow{K}_m\right)=\widetilde{\psi}_F\left(\overrightarrow{k'}+\overrightarrow{k}_B\right) \quad . \tag{2.43}$$

On the other hand, (2.41) can be inverted as

$$\psi\left(\overrightarrow{x}\right)\exp\left(-i\overrightarrow{k}_B\cdot\overrightarrow{x}\right)=\sum_{\overrightarrow{K}_m}\widetilde{\psi}\left(\overrightarrow{K}_m\right)\exp\left(i\overrightarrow{K}_m\cdot\overrightarrow{x}\right) \tag{2.44}$$

so that

$$\psi\left(\overrightarrow{x}+\overrightarrow{R}_n\right)=\exp\left(i\overrightarrow{k}_B\cdot\overrightarrow{R}_n\right)\psi\left(\overrightarrow{x}\right) \tag{2.45}$$

which concerns the three-dimensional form of Bloch's theorem (see Ashcroft and Mermin (1976)).

Inserting $\overrightarrow{k}_B=0$, one sees that (2.43) exhibits the limiting form

$$\widetilde{\psi}_F\left(\overrightarrow{k'}\right)=\widetilde{\psi}_F\left(\overrightarrow{k'}+\overrightarrow{K}_m\right) \tag{2.46}$$

which has been invoked in the study of the $1D$ harmonic oscillator on the lattice by Chabaud et al (1986) as

$$\widetilde{\psi}_F\left(k'\right)=\widetilde{\psi}_F\left(k'+\frac{2\pi}{a}\right) \quad . \tag{2.47}$$

Related details are presented in section 4.1. The Bloch-theorem in N space dimensions is also expressed usually as

$$\psi_{\overrightarrow{k}}\left(\overrightarrow{x}+\overrightarrow{a}\right)=\exp\left(i\overrightarrow{k}\cdot\overrightarrow{a}\right)\psi_{\overrightarrow{k}}\left(\overrightarrow{x}\right) \tag{2.48}$$

in accord with (2.45), where

$$\overrightarrow{k}\cdot\overrightarrow{a}=\sum_{i=1}^{N}\theta_i \tag{2.49}$$

and where $\theta_i = k_i a_i \in [-\pi, \pi]$. Other details concerning the Bloch-theorem may be found in textbooks (Ashcroft and Mermin (1976), Flügge (1971)).

2.4 Quasiperiodic potentials

Let us now assume that the potential can be expanded in a multiple Fourier series as

$$V(x)=\sum_{\{m_k\}} a_{m_1,\ldots,m_j}\exp\left(2\pi i\sum_{k=1}^{j} m_k\omega_k x\right) \tag{2.50}$$

where the m_k's are integers. If the frequency ω_k can be expressed by a quotient of two prime integers P_k and Q_k:

$$\omega_k=\frac{P_k}{Q_k} \tag{2.51}$$

then $V(x)$ exhibits j periods for which

$$V(x+Q_k)=V(x) \tag{2.52}$$

where $k = 1, 2, \ldots, j$. However, it may happen that the ω_k's are incommensurate. This means that for any set of rational numbers $\{r_k\}$ the equation (Besicovitch (1932), Romerio (1971)) the equation

$$\sum_{k=1}^{j} r_k\omega_k=0 \tag{2.53}$$

is able to be fulfilled for all k-values, if $r_k = 0$ only. This is e.g. the case if at least one of the frequencies is expressed by an irrational number. Then, one says that the potential is quasiperiodic, in which case the spectrum becomes an infinite Cantor-set (Ostlund and Pandit (1984), Simon (1982)). However, there are certain circumstances under which generalized versions of the Bloch-theorem may be applied as shown by Ostlund et al (1983a). Moreover, the periodic behavior is restored as soon as one resorts to rational approximations of irrationals. So far irrationals have also been approached via $P_k \rightarrow \infty$, and $Q_k \rightarrow \infty$, but for this purpose continued fractions can also be considered (Azbel (1979), Aubry and Andre (1980), Sokoloff (1981), Thouless (1983a), Wilkinson and Austin (1994)). Nevertheless, a problem remains, namely the question of whether an irrational number can be discriminated in practice from a rational approximation, or not. Indeed, in order to reproduce in practice irrational frequencies an infinite experimental resolution would be necessary, but this latter point looks rather questionable. On the other hand there are fractal-like experimental realizations which can be viewed as competition effects between rational and irrational frequencies, in which case successive resolutions with increasing accuracy are in order.

It is also clear that potentials on a discrete space for which the period is incommensurate with the spacing of the underlying lattice are able to produce nontrivial manifestations. Choosing the dimensionless lattice spacing to be unity, we then have to consider potentials for which

$$V(x + Q_i) = V(x) \tag{2.54}$$

where now Q_i is irrational. In particular, we can then consider the cosine potential

$$V(x) = V_0 \cos\left(2\pi \frac{P_1}{Q_i} x + \delta\right) \tag{2.55}$$

or so called Maryland-model (see also section 4.5)

$$V(x) = V_0 \tan\left(\pi \frac{P_1}{Q_i} x + \delta\right) \quad . \tag{2.56}$$

Furthermore, let us assume that $V(x)$ provides the diagonal potential of an $1D$ tight binding model with nearest-neighbor (NN) hopping (see also chapters 6 and 7). Using (2.56) it has been found (interestingly enough), that the spectrum is discrete for irrational Q_i values, such that all the states are normalizable (see Grempel et al (1982)). Accordingly, one deals with localized states, which is reminiscent to the well-known influence of random potentials (Anderson (1958)). This shows that quasiperiodic potentials are able to produce effects which are similar to the ones characterizing disordered systems, i.e. systems governed by diagonal random potentials specified above. In contradistinction, one obtains Q_1 bands as $Q_i \to Q_1$ approaches an integer, in which case the wavefunctions become increasingly extended. It is also been found that in the case of dissipative systems the transition from quasiperiodicity to chaos occurs both in a continuous and universal manner (Ostlund et al (1983b)).

In the sequel we shall confine, however, our attention on rational values of the commensurability parameter, with the understanding that irrationals are able to be handled as special limiting cases.

2.5 A shorthand presentation of the elliptic Lamè-equation

Further mathematical details concerning periodic Schrödinger equations (Simon (1985), Dyakin and Petrukhnovskii (1986)) are worthy of being mentioned. So there is an increasing interest in the application of non-unitary $SU(1,1)$- and $SU(2)$-representations to the band-energy description of Scarf- and Lamè-Hamiltonians, respectively (Li and Kusnezov (2000a)). In this context, we would like to say that the usual elliptic Lamè-equation

$$\frac{d^2\psi}{dx^2} + \kappa^2 l\,(l+1)\, sn^2 x \psi\,(x) = E\psi\,(x) \tag{2.57}$$

provides interesting applications in physics ranging from solitons to exactly solvable models (Li et al (2000b), Finkel et al (2000)). Here $snx = sn\,(x\,|\,\kappa\,)$ is the Jacobi elliptic sine function with modulus $0 \le \kappa \le 1$, whereas l is a real parameter playing the role of the quantum number of the angular momentum. Along the real axis sn^2x is periodic with period $2K$, where

$$K = \frac{\pi}{2} F\left(\frac{1}{2}, \frac{1}{2}, 1; \kappa^2\right) \tag{2.58}$$

denotes the complete elliptic integral. This equation can be converted into other systems, like Pöschl-Teller (Li and Kusnezov (2000a)) and Calogero-Moser (Enolskii and Eilbeck (1995)) systems, by resorting to suitable coordinate transformations and κ-selections. In this context, several classes of inter-related elliptic potentials and corresponding orthogonal polynomials have been discussed recently by Ganguly (2002). In addition, there are Scarf- and Mathieu-equation limits (Li et al (2000b)).

2.6 Quantum dot potentials

Nanoscaled systems such as quantum dots, nanorings, the 2D electron gas or the nanoconductors are multiparticle structures characterized by a reasonable small number of particles. Besides the single particle dynamics we then have to account for additional effects such as electron-electron interactions and the scattering of impurities. In addition, spin-orbit interactions can also be included. One recognizes that a complete description incorporating all these effects is computationally quite involved if not impossible. However, interactions referred to above may be less important in self assembled GaAs/GaAlAs-heterojunctions (Bastard (1992)), when the electrons occupy the lowest Landau-band only (Chakraborty et al (1994)). Under such conditions single particle electronic spectra are able to provide quite successfully useful information concerning thermodynamic and transport properties.

In this context there are reasons to say that the oscillator system is a useful tool to describe 2D- and 3D quantum dots (Simonin et al (2004)). The same concerns electrodynamic traps for the confinement of charged particles (Wuerker et al (1959), Wineland et al (1987)). The internal crystalline structure of quantum dots can be accounted for by resorting to the effective mass m^*, but descriptions on the discrete space remain desirable. To this aim discrete versions of the harmonic oscillator are presented in sections 4.1 and 5.6. A conceivable candidate for further developments, namely the q-deformed oscillator (Biedenharn (1989), Bonatsos et al (1994)), could also be proposed.

Of a special interest is the Fock-Darwin system, i.e. the single particle 2D parabolic quantum dot potential $kr_0^2/2$ supplemented by influence of a magnetic field $\overrightarrow{B}$ perpendicular to the plane of the dot. There is $r_0^2 = x^2 + y^2$, $k = m_0\omega_0^2/2$ and $q_e = -e < 0$, whereas the vector potential is selected in the symmetric gauge. This yields the Schrödinger-equation

$$\left[\frac{1}{2m_0}\left(-i\hbar\frac{\partial}{\partial x} - \frac{e}{2c}Bx\right)^2 + \frac{1}{2m_0}\left(-i\hbar\frac{\partial}{\partial y} + \frac{e}{2c}By\right)^2 + \right. \tag{2.59}$$

$$\left. + \frac{1}{2}kr_0^2\right]\Psi\left(\overrightarrow{x}\right) = E\Psi\left(\overrightarrow{x}\right) \quad ,$$

which can be rewritten equivalently as

$$\mathcal{H}_{FD}\Psi\left(\overrightarrow{x}\right) = \left(-\frac{\hbar^2}{2m_0}\Delta_2 + \frac{m_0}{2}\omega^2 r_0^2 + \frac{1}{2}\omega_B L_z\right)\Psi\left(\overrightarrow{x}\right) = E\Psi\left(\overrightarrow{x}\right) \tag{2.60}$$

by resorting to the third component of the angular momentum operator. There is

$$\omega = \left(\omega_0^2 + \frac{1}{4}\omega_B^2\right)^{1/2} = \omega_0 \left(1 + b^2\right)^{1/2} \quad (2.61)$$

where

$$\omega_B = \frac{eB}{m_0 c} = 2b\omega_0 \quad (2.62)$$

denotes the cyclotron frequency, while b is a transformation parameter. Using planar polar coordinates then gives the radial equation

$$-\frac{\hbar^2}{2m_0}\left(\frac{d^2}{dr_0^2} + \frac{1}{r_0}\frac{d}{dr_0}\right) F(r_0) + \quad (2.63)$$

$$+\left(\frac{m_\varphi^2}{2m_0 r_0^2} + \frac{m_0}{2}\omega^2 r_0^2\right) F(r_0) = \left(E - \frac{\hbar}{2} m_\varphi \omega_B\right) F(r_0)$$

where $m_\varphi = 0, \pm 1, \pm 2, \ldots$ is the magnetic quantum number. So all that needs is the exact solution of the two dimensional isotropic oscillator (see e.g. Kostelecky et al (1996)). Accordingly, the normalized wavefunction is given by

$$\Psi_n(r_0, \varphi; m_\varphi) = \left(\frac{2n_r!}{l_\omega\left(|m_\varphi| + n_r\right)!}\right)^{1/2} \frac{\exp(im_\varphi \varphi)}{\sqrt{2\pi}} F_n(r_0; |m_\varphi|) \quad (2.64)$$

where

$$F_n(r_0; |m_\varphi|) = \exp\left(-\frac{1}{2}\left(\frac{r_0}{l_\omega}\right)^2\right)\left(\frac{r_0}{l_\omega}\right)^{|m_\varphi|} L_n^{(|m_\varphi|)}\left(\frac{r_0^2}{l_\omega^2}\right) \quad . \quad (2.65)$$

One has $n = n_r$ and

$$l_\omega = \left(\frac{\hbar}{m_0 \omega}\right)^{1/2} \quad (2.66)$$

where $n_r = 0, 1, 2, \ldots$ denotes the radial quantum number, whereas $L_n^{(\alpha)}(x)$ stands for the Laguerre-polynomial. The corresponding energy reads

$$E = E_n(m_\varphi) = \hbar\omega_0 \left[\sqrt{1+b^2}\left(|m_\varphi| + 2n_r + 1\right) + bm_\varphi\right] \tag{2.67}$$

which proceeds in accord with (2.61) and (2.63). The dimensionless radial coordinate, say

$$r = \frac{r_0}{l_\omega} \tag{2.68}$$

can also be used, in which case (2.63) exhibits the typical form

$$\left[r^2\frac{d^2}{dr^2} + r\frac{d}{dr} + 2\left(|m_\varphi| + 2n_r + 1\right)r^2 - r^4\right]F(r) = m_\varphi^2 F(r) \quad . \tag{2.69}$$

So far, the wavefunction (2.64) obeys the periodic boundary condition

$$\Psi_n\left(r_0, \varphi; m_\varphi\right) = \Psi_n\left(r_0, \varphi + 2\pi; m_\varphi\right) \tag{2.70}$$

but an additional Aharonov-Bohm (AB) potential like (Aharonov and Bohm (1959))

$$A_\varphi = \frac{\Phi_A}{2\pi r_0}, \quad A_{r_0} = A_z = 0 \tag{2.71}$$

can also be inserted into (2.60). Here Φ_A stands for the thin magnetic flux confined along the Oz-axis. We can put, for convenience, $\Phi_A = \beta_A\Phi_0$, where $\Phi_0 = hc/e$ denotes the magnetic flux quantum, while β_A is a commensurability parameter. This amounts to replace m_φ by

$$m_\varphi \to m_\varphi + \beta_A \tag{2.72}$$

so that (2.70) becomes

$$\Psi_n\left(r_0, \varphi + 2\pi; m_A\right) = \exp(i2\pi\beta_A)\Psi_n\left(r_0, \varphi; m_A\right) \quad . \tag{2.73}$$

One remarks that (2.73) looks like the 1D Bloch-theorem (2.10), which opens the way to perform the identification

$$ka = 2\pi\beta_A \quad . \tag{2.74}$$

Now the lattice spacing a has to be viewed as a length scale which is proportional to a circumference.

On the other hand the velocity can be established as

$$v = \frac{1}{\hbar}\frac{dE}{dk} \tag{2.75}$$

where $E = E(k)$ is the related band energy, which is reminiscent to the parabolic energy dispersion law of the free particle. The corresponding current contribution is then given by (see also Chakraborty et al (1994))

$$I = -\frac{ev}{a} = -c\frac{dE}{d\Phi_A} \tag{2.76}$$

in accord with (2.74), which has the meaning of an equilibrium property at $T = 0$. This equation serves to the derivation of the persistent current by identifying E with the energies of quantum ring potentials, as we shall see later. The contribution of scattering states to the persistent current has also been established (Akkermans et al (1991)).

It should also be mentioned that inserting the cyclotron frequency (2.62) into (2.66) instead of ω yields the so called magnetic length

$$l_B = \sqrt{\frac{\hbar c}{eB}} \tag{2.77}$$

which serves as a typical length scale for systems threaded by a magnetic field. Competition effects between a and l_B would then be involved for systems on lattices pierced by a magnetic flux.

2.7 Quantum ring potentials

The displaced oscillator potential (Chakraborty et al (1994))

$$V_R(r_0) = \frac{1}{2}k(r_0 - R_0)^2 \tag{2.78}$$

is a useful candidate to the theoretical description of narrow nanoscopic semiconductor self-assembled rings for which the annular geometry effects are negligible. This potential exhibits a minimum at the ring radius $r_0 = R_0$. However, the quantum ring potential (2.78) is not exactly solvable, so that we have to resort to numerical or approximation methods. In contradistinction, the centrifugal "core" potential (Tan et al (1999))

$$V_R^{(c)}(r_0) = \frac{1}{2}k\left(r_0 - \frac{R_0^2}{r_0}\right)^2 \tag{2.79}$$

can be exactly solved. This proceeds by inserting

$$m_R = \left(m_\varphi^2 + \frac{m_0^2 \omega_0^2}{\hbar^2} R_0^4 \right)^{1/2} \tag{2.80}$$

instead of $|m_\varphi|$ in (2.64) and (2.65), with the understanding that $\exp(im_\varphi \varphi)$ in (2.64) is preserved as it stands. In order to reproduce data concerning actual GaAs-rings, selected numerical values like $m_0 = m^* = 0.067 m_e$, $R_0 \cong 800nm$ and $k \cong 4.44x10^{-5} meVnm^2$ have to be considered (Tan et al (1999)). The ring energy is then given by

$$E_R^{(\alpha)} = E_n(m_R) + kR_0^2 \tag{2.81}$$

which works in conjunction with (2.67) and (2.80), where α stands for the quantum numbers, i.e. $\alpha = \{m_\varphi, n_r\}$.

Moreover, the AB potential (2.71) piercing the ring can also be superimposed. The corresponding energy is then given by

$$\widetilde{E}_R^{(\alpha)} = \hbar\omega_0 \left[\sqrt{1+b^2} \left(\widetilde{m}_R + 2n_r + 1 \right) + b(m_\varphi + \beta_A) \right] + kR_0^2 \tag{2.82}$$

where

$$\widetilde{m}_R = \left((m_\varphi + \beta_A)^2 + \frac{m_0^2 \omega_0^2}{\hbar^2} R_0^4 \right)^{1/2} . \tag{2.83}$$

Azimuthal corrections to the ring potential like (Magnusdottir et al (1999))

$$V_R(r_0, \varphi) = \frac{1}{2} k_a \cos(2\varphi) r_0^2 \tag{2.84}$$

have also been considered.

2.8 Persistent currents and magnetizations

It can be readily verified that the derivative of the energy (2.82) with respect to the magnetic field is given by

$$\frac{\partial}{\partial B} \widetilde{E}_R^{(\alpha)} = \frac{e\hbar}{2m_0 c} \left[m_\varphi + \beta_A + \frac{\omega_B}{2\omega} (\widetilde{m}_R + 2n_r + 1) \right] \tag{2.85}$$

whereas the one with respect to Φ_A reads

$$\frac{\partial}{\partial \Phi_A} \widetilde{E}_R^{(\alpha)} = \frac{e\omega}{2\pi c \widetilde{m}_R} \left[m_\varphi + \beta_A + \frac{\omega_B}{2\omega} \widetilde{m}_R \right] \quad . \tag{2.86}$$

Fixing the number of electrons then gives the zero-temperature magnetization

$$M_\alpha = -\frac{\partial}{\partial B} \widetilde{E}_R^{(\alpha)} \tag{2.87}$$

so that the total magnetization is given by the sum over states as (Tan et al (1999))

$$M = \sum_\alpha M_\alpha|_{T=0} = \sum_\alpha M_\alpha \theta \left(E_F - E_\alpha \right) \tag{2.88}$$

where now $E_\alpha = \widetilde{E}_R^{(\alpha)}$ and where $\theta(x)$ denotes Heaviside's step function ($\theta(x) = 1$ for $x \geqslant 0$ and $\theta(x) = 0$ for $x < 0$). Furthermore, we can make the identification $E = \widetilde{E}_R^{(\alpha)}$, in which case the persistent current such as specified by (2.76) is given by

$$I_\alpha = -\frac{e\omega}{2\pi \widetilde{m}_R} \left[m_\varphi + \beta_A + \frac{\omega_B}{2\omega} \widetilde{m}_R \right] \tag{2.89}$$

the total current being given again by the summation over states:

$$I = \sum_\alpha I_\alpha|_{T=0} = \sum_\alpha I_\alpha \theta \left(E_F - E_\alpha \right) \tag{2.90}$$

It is understood that the summations in (2.88) and (2.90) have to be performed by accounting for energies less than the Fermi-energy. Note that such currents have been observed in GaAs/GaAlAs self-assembled rings (Mailly et al (1993)). Next one sees that

$$cM_\alpha = \pi l_\omega^2 \widetilde{m}_R I_\alpha - \frac{e\hbar\omega_B}{2m_0\omega} \left(n_r + \frac{1}{2} \right) \tag{2.91}$$

in which the first term from the r.h.s. is the magnetization produced by a current loop, whereas the second one represents a diamagnetic contribution. Applying the virial theorem, it can be easily verified that the expectation value of the squared radial coordinate with respect to the eigenfunction characterizing (2.82) is given by

$$< r_0^2 >= l_\omega^2(\widetilde{m}_R + 2n_r + 1) \tag{2.92}$$

in accord with (2.66). So the magnetization produced by the current loop can be rewritten as

$$cM_\alpha^{(loop)} = \pi < r_0^2 > \frac{\widetilde{m}_R}{\widetilde{m}_R + 2n_r + 1} \tag{2.93}$$

which approaches $\pi < r_0^2 >$ as $\widetilde{m}_R \rightarrow \infty$.

Choosing as an example free electrons on the 1D ring one obtains the number N_e of electrons as

$$N_e = \frac{2}{h}\int_0^L dx \int_0^{p_F} dp \tag{2.94}$$

where L denotes the ring circumference. The Fermi momentum $p_F = \hbar k_F$ is given by $E_F = p_F^2/2m_0$, as usual. Accordingly, there is

$$N_e = \frac{Lk_F}{\pi} \tag{2.95}$$

so that specifying the number of electrons amounts to consider a fixed value of the Fermi momentum.

Assuming that $T \neq 0$, we have to consider the canonical partition function

$$Z = \sum_\alpha \exp\left(-\frac{E_\alpha}{k_B T}\right) \tag{2.96}$$

for which the number of electrons is fixed. Inserting $E_\alpha = \widetilde{E}_R^{(\alpha)}$ into (2.96), one realizes that Z is an even periodic function of the AB magnetic flux Φ_A (Byers et al (1961)). The underlying background is the summation over the magnetic quantum number form $-\infty$ to ∞, which proceeds both for $T = 0$ and $T \neq 0$. This explains the appearance of AB-oscillations in the magnetic flux dependence of thermodynamic properties. So, the $T \neq 0$ counterpart of the $T = 0$ magnetization (2.88) is given by

$$M(\Phi, T) = -\frac{1}{Z}\sum_\alpha \frac{\partial}{\partial B} \exp\left(-\frac{E_\alpha}{k_B T}\right) \tag{2.97}$$

where Φ stands for the axial magnetic flux. Similarly, the temperature dependent persistent current reads

$$I(\Phi,T)=-\frac{1}{Z}\sum_{\alpha}\frac{\partial}{\partial\Phi}\exp\left(-\frac{E_{\alpha}}{k_BT}\right) \tag{2.98}$$

which is still a periodic, but an odd function of the magnetic flux Φ. We have to recognize that the derivation of temperature effects characterizing (2.95) and (2.96) is a lengthy job, but ingredients needed are presented in Appendix A in Cheung et al (1988). The influence of the electric field has also been considered (Barticevic et al (2002)). Thermodynamic properties relying on the grand canonical partition function are briefly discussed in section 8.6.

2.9 The derivation of the total persistent current for electrons on the 1D ring at $T = 0$

Let us consider as a next example free electrons on the planar 1D ring threaded by a magnetic flux Φ. The energies are given by Schrödinger equation

$$-\frac{\hbar^2}{2m_0R_0^2}\frac{d^2}{d\varphi^2}\psi(\varphi)=E\psi(\varphi) \tag{2.99}$$

working in conjunction with the AB-type periodic boundary condition

$$\psi(\varphi+2\pi)=\exp(i2\pi\beta)\,\psi(\varphi) \tag{2.100}$$

where $\beta=\Phi/\Phi_0$ and $\varphi\in[0,2\pi]$. Proceeding in this manner gives the energy and the eigenfunction as

$$E=E_n=\frac{\hbar^2}{2m_0R_0^2}(n+\beta)^2 \tag{2.101}$$

and

$$\psi=\psi_n(\varphi)=\frac{1}{\sqrt{2\pi}}\exp(i(\beta+n)\varphi) \tag{2.102}$$

respectively, where $n=0,\pm1,\pm2,\ldots$. The corresponding persistent current is then given by

$$I_n = -\frac{2}{N_e} I_F (n + \beta) \tag{2.103}$$

in accord with (2.76) and (2.95), where

$$I_F = \frac{e}{L} v_F \quad . \tag{2.104}$$

The circumference is denoted by $L = 2\pi R_0$, whereas

$$v_F = \frac{\hbar k_F}{m_0} \quad . \tag{2.105}$$

In order to derive the total persistent current at $T = 0$ we have to apply (2.90), which means that energies less than E_F have to be selected. This proceeds in terms of the inequalities (Cheung et al (1988))

$$-\frac{N_e}{2} \leqslant n + \beta \leqslant \frac{N_e}{2} \tag{2.106}$$

which produce in turn limitations which are sensitive to the parity of N_e. Performing summations over n then gives the total persistent current versus periodicity intervals as

$$I_+(\beta, N_e) = -2I_F(2\beta - 1)\frac{N_e + 1}{N_e} \tag{2.107}$$

and

$$I_-(\beta, N_e) = -4I_F\beta\frac{N_e + 1}{N_e} \tag{2.108}$$

if N_e is even ($\beta \in [0,1]$) and odd ($\beta \in [-1/2, 1/2]$), respectively. The corresponding β-intervals can be joined together as

$$\beta \in \left[-\frac{1}{2} + \frac{1 + (-1)^{N_e}}{4}, \frac{1}{2} + \frac{1 + (-1)^{N_e}}{4}\right] \tag{2.109}$$

where $(-1)^{N_e} = \pm 1$.

Resorting to Fourier series (see also (8.133)-(8.135)) leads to the generalized current

$$I_g(\beta, N_e) = \frac{2I_F(N_e + 1)}{\pi N_e} \sum_{l=1}^{\infty} \frac{\cos(l k_F L)}{l} \sin(2\pi l \beta) \tag{2.110}$$

which proceeds in accord with (2.94). One sees that

$$I_g(\beta+1, N_e) = I_g(\beta, N_e) \tag{2.111}$$

so that the oscillations characterizing the flux dependence of the total persistent current exhibit the unit period. Furthermore, one has the summation formula (see 1.448.1 in Gradshteyn and Ryzhik (1965))

$$\sum_{l=1}^{\infty} \frac{p^l \sin(lx)}{l} = \arctan\left(\frac{p\sin x}{1-p\cos x}\right) \tag{2.112}$$

where now $p = (-1)^{N_e} = \pm 1$. Accordingly

$$I_g(\beta, N_e) = \frac{2I_F(N_e+1)}{\pi N_e} \arctan\left(\frac{p\sin(2\pi\beta)}{1-p\cos(2\pi\beta)}\right) \tag{2.113}$$

which is valid for both even and odd N_e-values. Note that in order to derive (2.108) the integral

$$\int x\sin(ax)\,dx = -\frac{x}{a}\cos(ax) + \frac{1}{a^2}\sin(ax) \tag{2.114}$$

has been used.

The flux dependence of the dimensionless current

$$C = C_g(\beta, N_e) = \frac{I_g(\beta, N_e)}{I_F} \tag{2.115}$$

is displayed in figures 2.2 and 2.3 for $N_e = 3$ and $N_e = 4$, respectively. One sees that there is a π phase-difference in the oscillations characterizing these currents.

2.10 Circular currents

In order to establish alternatively circular currents in planar systems threaded by axial magnetic fields we can also resort to the current density

$$\vec{j} = \frac{i\hbar e}{2m_0}\left(\psi^*\nabla\psi - \psi\nabla\psi^*\right) - \frac{e^2}{m_0 c}\vec{A}\,|\psi|^2 = \tag{2.116}$$

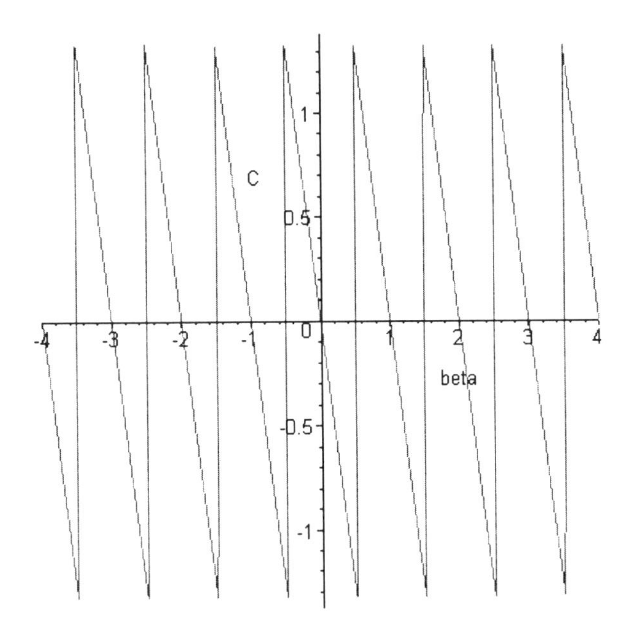

Fig. 2.2 **The flux dependence of $C_g(\beta, N_e)$ for $N_e = 3$. The sawtooth like oscillations displayed above are characterized by amplitudes having the magnitude order of 1.32.**

$$= -\left(\frac{\hbar e}{2m_0}\nabla \arg\psi + \frac{e^2}{m_0 c}\overrightarrow{A}\right)|\psi|^2$$

where $\overrightarrow{A}$ denotes the vector potential and $\psi = |\psi| \arg\psi$. Details concerning this rather special device have been presented before (Ziman (1964)). Choosing

$$\overrightarrow{A} = \frac{1}{2}\overrightarrow{B} \times \overrightarrow{r}_0 + \frac{\overrightarrow{\Phi}_A \times \overrightarrow{r}_0}{2\pi r_0^2} \tag{2.117}$$

we can account both for the influence of a transversal and homogeneous magnetic field B as well as for the axial AB -flux Φ_A. The tangential component characterizing (2.115) is then given by

$$j_\varphi = \frac{i\hbar e}{2m_0}\left(\psi^* \frac{1}{r_0}\frac{\partial\psi}{\partial\varphi} - \psi\frac{1}{r_0}\frac{\partial\psi^*}{\partial\varphi}\right) - \frac{e^2}{m_0 c}A_\varphi |\psi|^2 \tag{2.118}$$

where

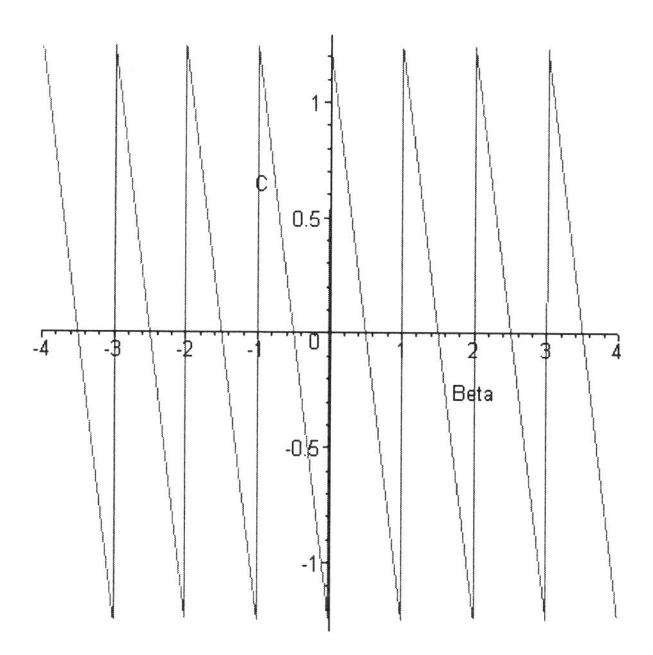

Fig. 2.3 **The flux dependence of $C_g(\beta, N_e)$ for $N_e = 4$. The sawtooth like oscillations displayed above are characterized by amplitudes having the magnitude order of 1.24.**

$$A_{\varphi} = \frac{Br_0}{2} + \frac{\Phi_A}{2\pi r_0} \tag{2.119}$$

Choosing

$$\psi = \frac{1}{\sqrt{2\pi}} \exp\left(im\varphi\right) \psi_{\alpha}\left(r_0\right) \tag{2.120}$$

which is reminiscent to (2.64), one finds the circular current as (see also Avishai et al (1993))

$$\widetilde{I}_{\alpha} = \int_0^{\infty} r_0 dr_0 j_{\varphi}^{(\alpha)}\left(r_0\right) \tag{2.121}$$

where

$$j_{\varphi}^{(\alpha)}\left(r_0\right) = -\frac{e\hbar}{m_0}\left|\psi_{\alpha}\left(r_0\right)\right|^2 \left(\frac{m+\beta_A}{r_0} + \frac{eB}{2c\hbar} r_0\right) \quad . \tag{2.122}$$

This serves as an alternative to (2.76).

The total current at $T = 0$ can then be established with the help of (2.90). Inserting

$$|\psi_0(r_0)|^2 = const.\delta(r_0 - R_0) \tag{2.123}$$

we have to remarks that (2.120) reproduces the persistent current characterizing the 1D ring such as given by (2.103) if $const. = \pi/L$ and $\beta = \beta_A$. This indicates that the present current can be rescaled on dimensional grounds as

$$\widetilde{I}_\alpha \to I_\alpha = \frac{\pi}{L}\widetilde{I}_\alpha \tag{2.124}$$

in which case the normalization of the wavefunction is safely preserved.

Chapter 3

Time Discretization Schemes

The unitary quantum mechanical evolution operator obeys the equation

$$\mathcal{H}\, U(t) = i\frac{\partial}{\partial t} U(t) \tag{3.1}$$

in simplified units for which $\hbar = 1$, which exhibits the well known solution

$$U(t) = \exp(-i\mathcal{H}t) \tag{3.2}$$

on the continuous time, provided that the Hamiltonian $\mathcal{H}$ is both Hermitian and independent of time. Considering a discrete space, amounts to replace $\mathcal{H}$ by a discretized counterpart, as we shall see later in chapter 4. Assuming a discrete time one should proceed similarly, now by replacing the usual time derivative $\partial/\partial t$ by a discrete derivative. We emphasize that this rule should be preserved even if the discrete time may look less fundamental than the the idea of a discrete space. This results in evolution equations proceeding on the discrete time, such as discussed in some more detail before (Caldirola (1976), Bender et al (1985), Bender and Dunne (1988), Lorente (1989)). Proofs have also been given that path integrals can be evaluated safely in terms of a countable discrete time (Lee (1983)). Assuming the time discretization $t = n\varepsilon$, where ε denotes a sufficiently small time scale and invoking, for convenience, (1.2), one then finds that (3.1) is subject to the discretization

$$\frac{i}{\varepsilon}(U_{n+1} - U_n) = \frac{1}{2}\mathcal{H}(U_{n+1} + U_n) \tag{3.3}$$

where $U(n\varepsilon) = U_n$. In addition, the U_n-wavefunction has been replaced by $(U_{n+1} + U_n)/2$, so as to make the discretization a little bit more elaborated.

Such equations will then be exercised in practice in the sequel. Conversely, taking the limits $\varepsilon \to 0$ and $n \to \infty$, but keeping $t = n\varepsilon$ fixed, one sees that (3.3) reproduces the continuous limit, i.e. (3.1), to $O\left(\varepsilon^2\right)$-order. Moreover, (3.3) exhibits the solution

$$U_n = \left(\frac{1 - \frac{i}{2}\varepsilon\mathcal{H}}{1 + \frac{i}{2}\varepsilon\mathcal{H}} \right)^n \tag{3.4}$$

where $U_0 = 1$, which reproduces again the continuous limit:

$$U_n \to \exp\left(-it\mathcal{H}\right) \quad , \tag{3.5}$$

as son as $\varepsilon \to 0$ and $n \to \infty$. Interesting results are also revealed by the q-deformation of the time evolution, as we shall see in the next sections. Such results are still open for further applications. Moreover, the time discretization can also be implemented in a close connection with the derivation of dynamic localization conditions (see (6.84)).

3.1 Discretized time evolutions of coordinate and momentum observables

An interesting class of time-dependent problems is the evolution of time-dependent observables in the Heisenberg representation. Indeed, one has

$$A\left(t\right) = U^{+}\left(t\right) AU\left(t\right) \tag{3.6}$$

where $A \equiv A\left(0\right)$, so that the evolution equation

$$i\frac{dA\left(t\right)}{dt} = \left[A\left(t\right), \mathcal{H}\right] \tag{3.7}$$

serves as a starting point for a subsequent discretization, too. In this context we shall discuss in particular the discretized time evolution of dimensionless and Hermitian coordinate (Q_n) and momentum (P_n) observables in terms of time-independent Hamiltonians of hyperbolic and elliptic type. The time discretized evolution equations can be introduced as

$$\frac{i}{\varepsilon}\left(Q_{n+1} - Q_n\right) = \frac{1}{2}\left[Q_{n+1} + Q_n, \mathcal{H}\right] \tag{3.8}$$

and

$$\frac{i}{\varepsilon}\left(P_{n+1}-P_{n}\right)=\frac{1}{2}\left[P_{n+1}+P_{n},\mathcal{H}\right] \tag{3.9}$$

proceeding in accord with (3.3), such that

$$\left[Q_{n},P_{n}\right]=i \tag{3.10}$$

irrespective of n.

3.2 Time independent Hamiltonians of hyperbolic type

Let us assume that the Hamiltonian $\mathcal{H}=\mathcal{H}(X)$ is an analytic function of X, where

$$X=Q_{0}P_{0}+P_{0}Q_{0}=Q_{n}P_{n}+P_{n}Q_{n} \tag{3.11}$$

is Hermitian an independent of time, i.e. of n. One finds

$$\left[Q_{n},X^{k}\right]=\left[\left(X+2i\right)^{k}-X^{k}\right]Q_{n} \tag{3.12}$$

in accord with (3.10), where k is an integer. Similarly, one has

$$\left[P_{n},X^{k}\right]=\left[\left(X-2i\right)^{k}-X^{k}\right]P_{n} \tag{3.13}$$

so that

$$\left[P_{n},\mathcal{H}(X)\right]=\left[\mathcal{H}(X-2i)-\mathcal{H}(X)\right]P_{n} \tag{3.14}$$

and

$$\left[Q_{n},\mathcal{H}(X)\right]=\left[\mathcal{H}(X+2i)-\mathcal{H}(X)\right]Q_{n}\quad . \tag{3.15}$$

Then (3.8) and (3.9) become

$$\frac{i}{\varepsilon}\left(Q_{n+1}-Q_{n}\right)=\frac{1}{2}\left(\mathcal{H}(X+2i)-\mathcal{H}(X)\right)\left(Q_{n+1}+Q_{n}\right) \tag{3.16}$$

and

$$\frac{i}{\varepsilon}\left(P_{n+1}-P_{n}\right)=\frac{1}{2}\left(\mathcal{H}(X-2i)-\mathcal{H}(X)\right)\left(P_{n+1}+P_{n}\right) \tag{3.17}$$

respectively. One sees that (3.16) and (3.17) are two-term recurrence relations, so that

$$Q_{n+1} = \frac{1 - \frac{i}{2}\varepsilon\left[\mathcal{H}(X+2i) - \mathcal{H}(X)\right]}{1 + \frac{i}{2}\varepsilon\left[\mathcal{H}(X+2i) - \mathcal{H}(X)\right]} Q_n \tag{3.18}$$

and

$$P_{n+1} = \frac{1 - \frac{i}{2}\varepsilon\left[\mathcal{H}(X-2i) - \mathcal{H}(X)\right]}{1 + \frac{i}{2}\varepsilon\left[\mathcal{H}(X-2i) - \mathcal{H}(X)\right]} P_n \tag{3.19}$$

respectively. One realizes, even at this stage of our calculations, that X is actually independent of n. Indeed, one has

$$Q^+_{n+1} = Q_{n+1} = Q_n \frac{1 + \frac{i}{2}\varepsilon\left[\mathcal{H}(X-2i) - \mathcal{H}(X)\right]}{1 - \frac{i}{2}\varepsilon\left[\mathcal{H}(X-2i) - \mathcal{H}(X)\right]} \tag{3.20}$$

so that

$$Q_{n+1}P_{n+1} = Q_n P_n \quad . \tag{3.21}$$

Concerning the product $P_n Q_n$ one proceeds similarly, which shows in turn that X is independent of n. Now it is obvious that the solution to the discretized time evolution of the coordinate is given by (see Lorente (1989))

$$Q_n = \left(\frac{1 - \frac{i}{2}\varepsilon\left[\mathcal{H}(X+2i) - \mathcal{H}(X)\right]}{1 + \frac{i}{2}\varepsilon\left[\mathcal{H}(X+2i) - \mathcal{H}(X)\right]}\right)^n Q_0 \quad . \tag{3.22}$$

The continuous limit of Q_n reads

$$Q_n \to \exp\left(-it\left(\mathcal{H}(X+2i) - \mathcal{H}(X)\right)\right) Q_0 \tag{3.23}$$

if $n \to \infty$ and $\varepsilon \to 0$, so that

$$Q_n \to \exp\left(it\mathcal{H}(X)\right) Q_0 \exp\left(-it\mathcal{H}(X)\right) \tag{3.24}$$

by virtue of (3.15). In the case of the momentum we have just to repeat the same steps.

3.3 Time independent Hamiltonians of elliptic type

In this case the Hamiltonian is an analytic function of the time-independent, dimensionless and Hermitian expression

$$y = P_n^2 + Q_n^2 \quad . \tag{3.25}$$

Now it is suitable to resort to boson annihilation and creation operators, in which case

$$a_n = \frac{1}{\sqrt{2}} \left(Q_n + iP_n \right) \tag{3.26}$$

and

$$a_n^+ = \frac{1}{\sqrt{2}} \left(Q_n - iP_n \right) \quad . \tag{3.27}$$

Accordingly, there is

$$y = a_n a_n^+ + a_n^+ a_n \tag{3.28}$$

irrespective of n. Now one has

$$\left[a_n, y^k \right] = \left[(y+2)^k - y^k \right] a_n \tag{3.29}$$

so that

$$\left[a_n^+, y^k \right] = -a_n^+ \left[(y+2)^k - y^k \right] \quad . \tag{3.30}$$

Inserting (3.26) and (3.27) into (3.8) and (3.9) yields

$$\frac{i}{2} \left(a_{n+1} - a_n \right) = \frac{1}{2} \left[a_{n+1} + a_n, \mathcal{H} \right] \tag{3.31}$$

and

$$\frac{i}{2} \left(a_{n+1}^+ - a_n^+ \right) = \frac{1}{2} \left[a_{n+1}^+ + a_n^+, \mathcal{H} \right] \tag{3.32}$$

as one might expect.

This time (3.31) produces the two-term recurrence relation (Lorente (1989))

$$a_{n+1} = \frac{1 - \frac{i}{2}\varepsilon\left[\mathcal{H}\left(y+2\right) - \mathcal{H}\left(y\right)\right]}{1 + \frac{i}{2}\varepsilon\left[\mathcal{H}\left(y+2\right) - \mathcal{H}\left(y\right)\right]} a_n \tag{3.33}$$

so that

$$a_{n+1}^{+} = a_n^{+} \frac{1 + \frac{i}{2}\varepsilon\left[\mathcal{H}\left(y+2\right) - \mathcal{H}\left(y\right)\right]}{1 - \frac{i}{2}\varepsilon\left[\mathcal{H}\left(y+2\right) - \mathcal{H}\left(y\right)\right]} \quad . \tag{3.34}$$

On sees again that the product $a_n^+ a_n$ is time-independent:

$$a_{n+1}^{+} a_{n+1} = a_n^{+} a_n \quad . \tag{3.35}$$

Moreover, one has the intertwining relation

$$a_n \mathcal{H}\left(y\right) = \mathcal{H}\left(y+2\right) a_n \tag{3.36}$$

by virtue of (3.29), which enables us to rewrite the r.h.s. of (3.33) by locating the a_n-factor to the left. This also means that

$$a_{n+1} a_{n+1}^{+} = a_n a_n^{+} \tag{3.37}$$

which confirms the n-independence of y.

3.4 The derivation of matrix elements

Let us introduce Fock states $|k\rangle$ for which

$$a_0 \left|k\right\rangle = \sqrt{k} \left|k-1\right\rangle \tag{3.38}$$

and

$$a_0^{+} \left|k\right\rangle = \sqrt{k+1} \left|k+1\right\rangle \tag{3.39}$$

as usual. Then

$$y \left|k\right\rangle = \left(2k+1\right) \left|k\right\rangle \tag{3.40}$$

so that

$$\left\langle l\right| \mathcal{H} \left|k\right\rangle = \mathcal{H}\left(2k+1\right) \delta_{l,k} \quad . \tag{3.41}$$

On the other hand,(3.33) yields the solution

$$a_n = \left(\frac{1 - \frac{i}{2}\varepsilon \left[\mathcal{H}\left(y+2\right) - \mathcal{H}\left(y\right) \right]}{1 + \frac{i}{2}\varepsilon \left[\mathcal{H}\left(y+2\right) - \mathcal{H}\left(y\right) \right]} \right)^n a_0 \tag{3.42}$$

in which case

$$\langle l|\, a_n \,|k\rangle = \left(\frac{1 - \frac{i}{2}\varepsilon \left[\mathcal{H}\left(2k+3\right) - \mathcal{H}\left(2k+1\right) \right]}{1 + \frac{i}{2}\varepsilon \left[\mathcal{H}\left(2k+3\right) - \mathcal{H}\left(2k+1\right) \right]} \right)^n \langle l|\, a_0 \,|k\rangle \quad . \tag{3.43}$$

But

$$\langle l|\, a_0 \,|k\rangle = \sqrt{k}\delta_{l,k-1} \tag{3.44}$$

so that both matrix elements $\langle l|\, a_n \,|k\rangle$ and $\langle l|\, a_n^+ \,|k\rangle$ are completely specified. Matrix elements of Q_n and P_n can also be easily obtained with the help of (3.26) and (3.27).

In the case of Hamiltonians of hyperbolic type, one needs the eigenvalues of the Hermitian X-operator such is given by (3.11). Using (3.26) and (3.27) gives

$$X = -i\left(a_0^2 - a_0^{+^2}\right) \tag{3.45}$$

but unfortunately, the Fock states are not eigenfunctions of this operator. We then have to introduce a differential realization like $a_0 \rightarrow d/d\zeta$ and $a_0 \rightarrow \zeta$ (Lorente (1989)). Then the X-eigenvalues, say 2ρ, are produced by the equation

$$-i\left(\frac{d^2}{d\zeta^2} - \zeta^2\right)\psi_\rho\left(\zeta\right) = 2\rho\psi_\rho\left(\zeta\right) \quad . \tag{3.46}$$

The corresponding eigenfunction gets expressed in terms of the parabolic cylinder function (see e.g. Gradsteyn and Ryzhik (1965)) as

$$\psi_\rho\left(\zeta\right) = D_{-i\rho-1/2}\left(\sqrt{2}\zeta\right) \tag{3.47}$$

so that the zero-time matrix elements are

$$\langle\rho|\, Q_0 \,|\rho\prime\rangle = \frac{1}{\sqrt{2}} \int_{-\infty}^{+\infty} \psi_\rho^*\left(\zeta\right)\left(\zeta + \frac{d}{d\rho}\right)\psi_{\rho'}\left(\zeta\right) d\zeta \tag{3.48}$$

and

$$\langle\rho|\, P_0\, |\rho\prime\rangle = \frac{i}{\sqrt{2}} \int_{-\infty}^{+\infty} \psi_\rho^*(\zeta)\left(\zeta - \frac{d}{d\rho}\right)\psi_{\rho'}(\zeta)\, d\zeta \quad . \tag{3.49}$$

The matrix element of the n-dependent coordinate is then given by

$$\langle\rho|\, Q_n\, |\rho\prime\rangle = \left(\frac{1 - \frac{i}{2}\varepsilon\left(\mathcal{H}(2\rho + 2i) - \mathcal{H}(2\rho)\right)}{1 + \frac{i}{2}\varepsilon\left(\mathcal{H}(2\rho + 2i) - \mathcal{H}(2\rho)\right)}\right)^n \langle\rho|\, Q_0\, |\rho\prime\rangle \tag{3.50}$$

in accord with (3.22). It is obvious that a similar relation is valid for the momentum P_n.

3.5 Finite difference Liouville-von Neumann equations and "elementary" time scales

A finite difference formulation of the Liouville-von Neumann equation for the density operator:

$$i\hbar\frac{\rho(t+\tau) - \rho(t-\tau)}{2\tau} = [H, \rho(t)] \tag{3.51}$$

has also been discussed by Bonifacio and Caldirola (1983a), where τ represents a characteristic time scale of the system. This corresponds to a symmetrization of the difference quotients introduced before by virtue of (1.1) and (1.2). A similar symmetrization has also been used in terms of the finite difference Schrödinger-equation (Bonifacio (1983b), Janussis (1984))

$$i\hbar\frac{\psi(t+\tau) - \psi(t-\tau)}{2\tau} = H\psi(t) \tag{3.52}$$

in which τ plays the role of an "elementary" time scale. A such time scale relies on several time-quantization ideas proposed before (see e.g. Pokrowski (1928) and Beck (1929)) and especially on the "chronon" hypothesis" forwarded by Caldirola (1976). Moreover, accounting for discrete time effects results in certain corrections to the relativistic mass energy formula (Wolf (1989)). Of course the "elementary" time can also be approached via $\tau = a/c$, where a plays the role of an "elementary" length.

Equation (3.51) can be easily solved in terms of the matrix elements

$$\rho_{nm}(t) = \langle n| \rho(t) |m\rangle \tag{3.53}$$

where $|n\rangle$ and $|m\rangle$ are non-degenerate energy eigenfunctions:

$$H|n\rangle = E_n|n\rangle \tag{3.54}$$

as usual. So, (3.51) becomes

$$\rho_{nm}(t+\tau) - \rho_{nm}(t-\tau) = -2i\omega_{nm}\tau\rho_{nm}(t) \tag{3.55}$$

where

$$\omega_{nm} = \frac{1}{\hbar}(E_n - E_m) \quad . \tag{3.56}$$

The solution of (3.55) reads

$$\rho_{nm}(t) = \rho_{nm}(0)\exp(-k_{nm}t) \tag{3.57}$$

where

$$\exp(-k_{nm}\tau) = -i\tau\omega_{nm} \pm \sqrt{1-\omega_{nm}^2\tau^2} \quad . \tag{3.58}$$

Putting

$$k_{nm} = \gamma_{nm} + i\upsilon_{nm} \tag{3.59}$$

one finds further details by accounting for the continuous limit

$$\rho_{nm}(t) = \rho_{nm}(0)\exp(-i\omega_{nm}t) \quad . \tag{3.60}$$

This shows that one should have $\gamma_{nm} \to 0$ and $\upsilon_{nm} \to \omega_{nm}$ as $\omega_{nm}\tau \to 0$. Next, $\exp(-k_{nm}t)$ should be bounded as $t \to \infty$, which means in turn that $\gamma_{nm} > 0$. In addition, we shall consider a rather reasonable ansatz, namely that k_{nm} should be a continuous function of $\omega_{nm}\tau$, which concerns especially the points $\omega_{nm}\tau = \pm 1$.

First let us assume that $|\omega_{nm}|\tau \geq 1$. Then

$$\upsilon_{nm} = \frac{\pi}{2\tau} sgn(\omega_{nm}\tau) \tag{3.61}$$

whereas

$$\gamma_{nm} = \frac{1}{\tau} \ln \left[|\omega_{nm}| \tau + \sqrt{\omega_{nm}^2 \tau^2 - 1} \right] \quad . \tag{3.62}$$

Complementarily, one obtains

$$\gamma_{nm} = 0 \tag{3.63}$$

and

$$\upsilon_{nm} = \frac{1}{\tau} \arcsin \left(|\omega_{nm}| \tau \right) \tag{3.64}$$

if $|\omega_{nm}| \tau \leq 1$.

One remarks that the influence of the characteristic time τ concerns not only the rescaling of the transition frequency ω_{nm}, but of the real and imaginary parts of k_{nm} ,too. We then have to realize that the discrete description produces richer structures than the continuous one and this is the reason why discretizations look both interesting and promising.

3.6 The q-exponential function approach to the q-deformation of time evolution

Equation (3.2) shows that in order to perform the q-deformation of the quantum-mechanical time evolution we have to introduce the q-exponential functions, say

$$\exp(\beta t) \rightarrow \exp_q(\beta t) \tag{3.65}$$

working in combination with appropriate q-derivatives (Ubriaco (1992)). Choosing the Jackson-derivative, we have to account for the substitution

$$\frac{\partial}{\partial t} \rightarrow \partial_t^{(q)} \tag{3.66}$$

which works in accord with (1.12). Then one deals with the q-deformed exponential functions like

$$\exp_q(x) = \sum_{k=0}^{\infty} \frac{x^k}{[[k]]_q!} \tag{3.67}$$

in which case

$$\partial_t^{(q)} \exp_q(\beta t) = \beta \exp_q(\beta t) \quad . \tag{3.68}$$

Another realization such as

$$Exp_q(x) = \sum_{k=0}^{\infty} \frac{q^{n(n-1)/2} x^k}{[[k]]_q!} \tag{3.69}$$

can also be considered, so that

$$\partial_t^{(q)} Exp_q(\beta t) = \beta Exp_q(\beta q t) \quad . \tag{3.70}$$

Moreover, the symmetrized q-derivative $\mathcal{D}_x^{(q)}$ can also be invoked. Then (3.67) and (3.68) get replaced by

$$e_q(x) = \sum_{k=0}^{\infty} \frac{x^k}{[k]_q!} \tag{3.71}$$

and

$$\mathcal{D}_t^{(q)} e_q(\beta t) = \beta e_q(\beta t) \tag{3.72}$$

respectively. Furthermore, the symmetrized counterpart of (3.70) reads

$$\mathcal{D}_z^{(q)} E_q(\beta t) = \beta E_q(\beta q t) \tag{3.73}$$

where now

$$E_q(x) = \sum_{k=0}^{\infty} \frac{q^{n(n-1)/2} x^k}{[k]_q!} \quad . \tag{3.74}$$

We are ready to remark that the q-exponential functions introduced above are able to exhibit non-classical properties by virtue of the inherent non-additivity behavior with respect to the argument. Indeed, one has

$$\exp_q(x+y) \neq \exp_q(x) \exp_q(y) \tag{3.75}$$

if $q \neq 1$, which can be easily verified by using the $a = 0$ version of (1.21):

$$\exp_q\left(\frac{x}{1-q}\right) = \prod_{k=0}^{\infty} \frac{1}{(1 - xq^k)} \quad . \tag{3.76}$$

Other details concerning q-exponential functions can be found in the literature (Nelson and Gartley (1994), MacAnally (1995)).

The next point of interest is the derivation of the q-deformed Baker-Campbell-Hausdorf formula (see e.g. Ubriaco (1992), Chung (1993))

$$\exp_q(A)\, B \exp_{1/q}(-A) = B + [A,B] + \frac{1}{[[2]]_q!}\left[A,[A,B]_q\right] + \tag{3.77}$$

$$+\frac{1}{[[3]]_q!}\left[A,\left[A,[A,B]_{q^2}\right]_q\right] + \ldots$$

in which the general term looks like

$$T_n^{(BCH)}(q) = \frac{1}{[[n]]_q!}\left[A,\left[A,\ldots[A,B]_{q^{n-1}}\right]_{q^{n-2}}\cdots\right]_q\Bigg] \quad . \tag{3.78}$$

The q-deformed commutator displayed above reads

$$[A,B]_{q^n} = AB - q^n BA \tag{3.79}$$

so that

$$\left[A,[A,B]_{q^n}\right]_{q^m} = \left[A,[A,B]_{q^m}\right]_{q^n} \quad . \tag{3.80}$$

This means in turn, that the ordering of q-commutators is irrelevant. Equation (3.77) can be verified by performing q-difference Taylor-series expansions of the function

$$f(\lambda) = \exp_q(\lambda A)\, B \exp_{1/q}(-\lambda A) \tag{3.81}$$

which works as follows

$$f(\lambda) = f(0) + \partial_\lambda^{(q)} f(\lambda)|_{\lambda=0} + \frac{1}{[[2]]_q!}\partial_\lambda^{(q)^2} f(\lambda)|_{\lambda=0} + \ldots \quad . \tag{3.82}$$

For this purpose we have to apply the derivatives

$$\partial_\lambda^{(q)} \exp_q(\lambda A) = A \exp_q(\lambda A) \tag{3.83}$$

and

$$\partial_\lambda^{(q)} \exp_{1/q}(-\lambda A) = -A \exp_{1/q}(-\lambda A q) \tag{3.84}$$

where

$$[[n]]_{1/q}! = \frac{[[n]]_q!}{q^{n(n-1)/2}} \quad . \tag{3.85}$$

Proceeding in conjunction with (1.13) yields the concrete results

$$f(0) = B \tag{3.86}$$

$$\partial_\lambda^{(q)} f(\lambda)|_{\lambda=0} = [A, B] \tag{3.87}$$

$$\partial_\lambda^{(q)^2} f(\lambda)|_{\lambda=0} = \left[A, [A, B]_q\right] \tag{3.88}$$

and similarly to higher orders. In particular, we can put $A = x$ and $B = 1$, in which case (3.77) becomes

$$\exp_q(x) \exp_{1/q}(-x) = 1 \quad . \tag{3.89}$$

This provides a useful solution to the inverse of the q-exponential function. Of course,(3.77) works also in terms of other q-exponential functions, but in this case we have to consider appropriate realizations of the q-commutator instead of (3.79).

Having obtained the main ingredients, one realizes that the q-deformation of (3.6) is an easy exercise. For this purpose we shall apply the q-deformation of the evolution operator as

$$U(t) \to U_q(t) = \exp_q(-\frac{i}{2}t\mathcal{H}) \exp_{1/q}(-\frac{i}{2}t\mathcal{H}) \tag{3.90}$$

where q is real, which satisfies unitarity requirements needed:

$$U_q^+(t) = U_q^{-1}(t) \tag{3.91}$$

by virtue of (3.89). Using (3.83) and (3.84) it can be easily verified that

$$i\partial_t^{(q)} U_q(t) = \frac{\mathcal{H}}{2}\left(U_q(t) + U_q(qt)\right) \tag{3.92}$$

which leads to

$$-i\partial_t^{(q)} U_q^+(t) = \frac{\mathcal{H}}{2}\left(U_q^{-1}(t) + U_q^{-1}(qt)\right) \tag{3.93}$$

by Hermitian conjugation. The related q-deformed time evolution of an observable is then given by

$$A_q(t) = U_q^+(t)A(0)U_q(t) \tag{3.94}$$

so that

$$\partial_t^{(q)} A_q(t) = \frac{i}{2}\left(\mathcal{H}A_q(t) - A_q(qt)\mathcal{H}\right) + \tag{3.95}$$

$$+\frac{i}{2}U_q^+(qt)[\mathcal{H}, A(0)]U_q(t)$$

which exhibits certain similarities with (3.8). On sees that (3.95) reproduces (3.7) as soon as $q \to 1$, as one might expect.

Using (3.77) leads to

$$\exp_{1/q}(\frac{i}{2}t\mathcal{H})\mathcal{A}(0)\exp_q(-\frac{i}{2}t\mathcal{H}) = A(0) + \frac{i}{2}t[\mathcal{H}, A(0)] + \tag{3.96}$$

$$+\left(\frac{it}{2}\right)^2 \frac{1}{[[2]]_{1/q}!}[\mathcal{H}, [\mathcal{H}, A(0)]_{1/q}] + \ldots \quad .$$

Next, one obtains

$$\exp_q(\frac{i}{2}t\mathcal{H})\mathcal{A}(0)\exp_{1/q}(-\frac{i}{2}t\mathcal{H}) = A(0) + \frac{i}{2}t[\mathcal{H}, A(0)] + \tag{3.97}$$

$$+\left(\frac{it}{2}\right)^2 \frac{1}{[[2]]_q!}[\mathcal{H}, [\mathcal{H}, A(0)]_q] + \ldots$$

with the understanding that similar formulae can be done for all non-zero terms from the r.h.s. of (3.96). This shows that $A_q(t)$ can be established in an explicit manner by resorting to underlying non-zero commutators.

3.7 Alternative realizations of discrete time evolutions and stationary solutions

Alternative realizations to the discrete time evolution are able to be provided by the q-deformed formulae written down above by applying the substitution

$$t \rightarrow q^n \tag{3.98}$$

in which one assumes again that q is real. Accordingly, the unitary time evolution operator becomes

$$U_q(t) \rightarrow U_q^{(n)} = \exp_q\left(-\frac{i}{2}q^n\mathcal{H}\right)\exp_{1/q}\left(-\frac{i}{2}q^n\mathcal{H}\right) \tag{3.99}$$

so that the unitarity condition becomes

$$\left(U_q^{(n)}\right)^+ U_q^{(n)} = 1 \quad . \tag{3.100}$$

Then the discrete counterparts of (3.92) and (3.95) are given by

$$\frac{i}{q^n(q-1)}\left(U_q^{(n+1)} - U_q^{(n)}\right) = \frac{\mathcal{H}}{2}\left(U_q^{(n+1)} + U_q^{(n)}\right) \tag{3.101}$$

which relies on (3.3), and

$$\frac{A_q^{(n+1)} - A_q^{(n)}}{q^n(q-1)} = \frac{i}{2}\left[\mathcal{H}A_q^{(n)} - A_q^{(n+1)}\mathcal{H}\right] + \frac{i}{2}\left(U_q^{(n+1)}\right)^+ \left[\mathcal{H}, A(0)\right] U_q^{(n)} \tag{3.102}$$

respectively, where

$$A_q^{(n)} = A_q(q^n) \quad . \tag{3.103}$$

Of course, stationary solutions to (3.52):

$$\psi(x) = \psi(\overrightarrow{x}, t) = \exp(-i\omega t)\varphi(\overrightarrow{x}) \tag{3.104}$$

can also be easily done. Inserting (3.104) into (3.52) yields

$$E = \frac{\hbar}{\tau}\sin\omega\tau \tag{3.105}$$

where

$$\mathcal{H}\varphi\left(\vec{x}\right) = E\varphi\left(\vec{x}\right) \quad . \tag{3.106}$$

Then (3.104) can be rewritten as (see also Caldirola (1976))

$$\psi\left(\vec{x}, t\right) = \exp\left(-i\frac{t}{\tau}\arcsin\left(\frac{E\tau}{\hbar}\right)\right)\varphi_E\left(\vec{x}\right) \tag{3.107}$$

where $\varphi_E\left(\vec{x}\right) = \varphi\left(\vec{x}\right)$. However, this time the energy is subject to the upper bound

$$|E| \leq E^{(\max)} = \frac{\hbar}{\tau} \tag{3.108}$$

which is an immediate consequence of (3.105). We have to anticipate that the discrete analog of the harmonic oscillator is characterized by the maximum energy

$$E^{(\max)} = \hbar\omega_0\left(N_0 + \frac{1}{2}\right) \tag{3.109}$$

where now $\hbar\omega_0 \neq 1$, which proceeds in accord with (5.39) and (5.40). The corresponding "chronon" is then given by

$$\tau = \frac{T_0}{\pi\left(2N_0 + 1\right)} \tag{3.110}$$

where $T_0 = 2\pi/\omega_0$ denotes the period of the oscillator and where N_0 is a fixed integer.

Chapter 4

Discrete Schrödinger Equations. Typical Examples

Schrödinger equations on the discrete space can be readily established by resorting to the discrete left- and right-hand derivatives written down before. A first realization can be done by replacing the usual second-derivative by its discrete counterpart $\nabla\Delta = \Delta\nabla$. This leads to the $1D$ discrete Schrödinger-equation

$$-\nabla\Delta\psi(x) + V_1(x)\psi(x) = E_1\psi(x) \tag{4.1}$$

where $V_1(x)$ stands for the usual potential and where x denotes the dimensionless discrete variable. This equation can be rewritten equivalently as

$$-\psi(x+1) - \psi(x-1) + V_1(x)\psi(x) = E\psi(x) \tag{4.2}$$

in accord with (1.5), where $E = E_1 - 2$. However, in many cases one deals with a slightly modified form:

$$\mathcal{H}\psi(x) = \psi(x+1) + \psi(x-1) + V(x)\psi(x) = E^{'}\psi(x) \tag{4.3}$$

where $V(x) = -V_1(x)$ and $E^{'} = -E$. Hopping effects can also be included into (4.2) and (4.3) by multiplying ψ_{n+1} and ψ_{n-1} with hopping matrix elements needed. We have to remark that the Hamiltonian characterizing (4.2) can be rewritten in terms of translations as

$$\mathcal{H} = -\left[\exp\left(\frac{\partial}{\partial x}\right) + \exp\left(-\frac{\partial}{\partial x}\right)\right] + V_1(x) \tag{4.4}$$

so that

$$\mathcal{H} = -2\cosh\left(\frac{\partial}{\partial x}\right) + V_1(x) \tag{4.5}$$

or

$$\mathcal{H} = -2\cos p_{op} + V_1(x) \tag{4.6}$$

where $p_{op} = -i\partial/\partial x$ denotes again the dimensionless momentum operator. Starting e.g. with (4.6) we can then account for the discrete coordinate at the end of calculations. It is also clear that (4.5) can be easily generalized to N space-dimensions as

$$\mathcal{H} = -2\sum_{j=1}^{N}\cosh\left(\frac{\partial}{\partial x_j}\right) + V_1(\vec{x}) \tag{4.7}$$

where $\vec{x} = (x_1, x_2, \ldots, x_N)$. The same concerns, of course, (4.6). Such equations are produced specifically by tight binding models to the description of electrons on square, or other kind of lattices under the influence of external fields. However, well-known examples from usual quantum mechanics, such as the harmonic oscillator (Chabaud et al (1986)), the Coulomb -potential (Gallinar (1984), Kvitsinsky (1992)), or the linear potential (Gallinar and Mattis (1985)) are worthy of being mentioned, too. Moreover, the so-called Maryland-potential has also been discussed in some more detail (Kvitsinsky (1994), Simon (1985) and Grempel et al (1982)). Such studies reveal quite interesting manifestations of space-discreteness. The infinite square well has been discussed recently by Boykin and Klimeck (2004) by accounting in some more detail for perfect confinement boundary conditions. However a simplified description can also be readily done by proceeding in a close connection with the usual description.

4.1 The isotropic harmonic oscillator on the lattice

The isotropic quantum harmonic oscillator on a cubic lattice is described by the Hamiltonian

$$\mathcal{H} = -2t_H\sum_{j=1}^{3}\cosh\left(\frac{\partial}{\partial x_j}\right) + \frac{k}{2}\sum_{j=1}^{3}x_j^2 \tag{4.8}$$

where t_H is responsible for site independent hopping effects. The related eigenvalue equation

$$\mathcal{H}\Psi\left(\overrightarrow{x}\right) = \mathcal{E}\Psi\left(\overrightarrow{x}\right) \tag{4.9}$$

is of a special interest in semiconductor physics. The present wavefunction separates as

$$\Psi\left(\overrightarrow{x}\right) = \prod_{j=1}^{3} \psi\left(x_j\right) \tag{4.10}$$

in accord with the usual description. Accordingly, the energy reads

$$\mathcal{E} = \sum_{j=1}^{3} E_j \tag{4.11}$$

where the quotations are selfconsistently understood. So, (4.9) gets solved in terms of the $1D$ eigenvalue problem

$$-t_H \cos p_{op}\psi\left(x\right) + \frac{k}{2}x^2\psi\left(x\right) = E\psi\left(x\right) \quad . \tag{4.12}$$

Resorting, for convenience, to the Fourier-decomposition

$$\psi\left(x\right) = \int_{-\infty}^{+\infty} f\left(s\right)\exp\left(isx\right)ds \tag{4.13}$$

leads to the s-momentum representation equation

$$-2t_H \cos\left(s\right) f\left(s\right) - \frac{k}{2}\frac{d^2 f}{ds^2} = Ef(s) \tag{4.14}$$

with the understanding that a similar result would obtained by using summations over the Brillouin zone instead of (4.13). Both $\psi\left(x\right)$ and $f\left(s\right)$-functions are assumed to be square integrable. One sees that (4.14) exhibits the canonical form of the Mathieu-equation (Abramowitz and Stegun (1972))

$$\frac{d^2 f}{du^2} + \left(a_M - 2q_M \cos 2u\right) f = 0 \quad , \tag{4.15}$$

via $s = 2u$,

$$q_M = -\frac{8A_H}{k} \tag{4.16}$$

and

$$a_M = \frac{8E}{k} \quad . \tag{4.17}$$

On the other hand, the reciprocal lattice (Ashcroft and Mermin (1976)), is characterized by the period $T_p = 2\pi$, so that the admissible solutions should satisfy the periodicity condition

$$f(u) = f(u + \pi) \tag{4.18}$$

which works in accord with (2.47). At this point, we have to remind that there exists a countable infinite set of real and distinct characteristic a_M-values, say

$$a_M = a_{2\rho}(q_M) \tag{4.19}$$

where $\rho = 0, 1, 2, \ldots,$ and

$$a_M = b_{2\rho}(q_M) \tag{4.20}$$

for $\rho = 1, 2, \ldots,$ yielding even and odd periodic solutions of period π, respectively. Accordingly, one has the Mathieu-wavefunctions

$$f(u) = N_{2r}^{(c)} ce_{2\rho}(u; q_M) = \sum_{n=0}^{\infty} A_{2n} \cos(2nu) \tag{4.21}$$

and

$$f(u) = N_{2r}^{(s)} se_{2\rho}(u; q_M) = \sum_{n=0}^{\infty} B_{2n} \sin(2nu) \tag{4.22}$$

where the coefficients A_{2n} and B_{2n} are able to be established with the help of three-term recurrence relations. The corresponding normalization constants are denoted by $N_{2r}^{(c)}$ and $N_{2r}^{(s)}$.

Proceeding in this manner one finds the energy-eigenvalues

$$E = E_{\rho}^{(a)}(q_M) = \frac{k}{8} a_{2\rho}(q_M) \tag{4.23}$$

and

$$E = E_{\rho}^{(b)}(q_M) = \frac{k}{8} b_{2\rho}(q_M) \tag{4.24}$$

respectively, for which power series in q_M can be derived. So, one has the limit

$$E \rightarrow \frac{k}{2} \rho^2 \tag{4.25}$$

if $q_M \rightarrow 0$, which differs drastically from the behavior characterizing the usual harmonic oscillator. However, there is

$$a_{2\rho} \cong b_{2\rho+1} \cong -2|q_M| + 2(2\rho+1)\sqrt{|q_M|} \tag{4.26}$$

if $|q_M| \gg 1$. Then

$$E_{\rho}^{(a)} \cong -2A_H + \left(\rho + \frac{1}{2}\right) \omega_0 \sqrt{t_H} \tag{4.27}$$

where $k \equiv \omega_0^2/2$, which reproduces precisely the spectrum of the usual harmonic oscillator if $t_H = 1$. The real-space counterparts of the above wavefunctions have also been discussed by Chabaud et al (1986). In other words, the harmonic oscillator on the lattice produces energy eigenvalues depending in a nontrivial manner on the q_M-parameter, which goes beyond the capabilities of the usual harmonic oscillator on the usual line. Note that the discrete harmonic oscillator has also been revisited recently by Aunola (2003), by resorting again to the Mathieu-equation, which results in a new solution expressed in terms of generalized Hermite-polynomials.

4.2 Hopping particle in a linear potential

The discrete Schrödinger equation describing a particle with charge q_e and hopping with a constant matrix element t_H under the influence of an applied electric field $\overrightarrow{\mathcal{E}}_{el} = (\mathcal{E}_{el}, 0, 0)$ is given by (Gallinar and Mattis (1985))

$$-t_H(\psi_{n+1} + \psi_{n-1}) + U_n \psi_n = E \psi_n \tag{4.28}$$

where

$$U_n = \mathcal{F}n \tag{4.29}$$

stands for a linear potential, while

$$\mathcal{F} = -q_e \mathcal{E}_{el} > 0 \quad . \tag{4.30}$$

One realizes immediately that the discrete Hamiltonian characterizing (4.28) is

$$\mathcal{H} = -2t_H \cosh \frac{\partial}{\partial n} + \mathcal{F}n \tag{4.31}$$

where $x \equiv n \in \mathbb{Z}$. The continuous limit of (4.31) reads

$$\mathcal{H}_c = -t_H \frac{d^2}{dx^2} + \mathcal{F}x - 2t_H \tag{4.32}$$

in which case the wavefunctions are expressed in terms of Airy-functions (Abramowitz and Stegun (1972))

$$A_{i_\pm}(s) \equiv \frac{\sqrt{s}}{3} J_{\pm 1/3}\left(\frac{2}{3}s^{3/2}\right) \tag{4.33}$$

as

$$\psi_n = \psi(s) = A_{i_\pm}(s) + A_{i_\mp}(s) \tag{4.34}$$

where

$$s = \frac{1}{\mathcal{F}^{2/3} t_H^{1/3}} (E + 2t_H - \mathcal{F}x) \quad . \tag{4.35}$$

However, such wavefunctions are not square integrable, which indicates that there are no discrete energy levels.

On the other hand, the Bessel-functions of the first kind, say $J_\nu(z_0)$, satisfy the recurrence relation

$$J_{\nu-1}(z_0) + J_{\nu+1}(z_0) = \frac{2\nu}{z} J_\nu(z_0) \tag{4.36}$$

in which case (4.28) exhibits the solution

$$\psi_n = \psi_n(z_0) = J_{n-w}(z_0) \tag{4.37}$$

where

$$z_0 = 2\frac{t_H}{\mathcal{F}} \tag{4.38}$$

and

$$\nu = n - w = n - \frac{E}{\mathcal{F}} \quad . \tag{4.39}$$

For this purpose (4.28) has been converted into the momentum representation by resorting specifically to Fourier transforms. In contradistinction to the Bessel functions of the second kind, the $J_\nu(z_0)$ functions considered above get favored as they are normalizable on the ν -space. So far the quotient $E/\mathcal{F}$ is not quantized in an explicit manner, but this question will be clarified in section 6.12 by using summations over the first Brillouin zone instead of Fourier transforms.

Next let us look, however, for a suitable boundary condition. For this purpose we can insert an infinite high barrier into (4.28), such that U_n is replaced by (Gallinar and Mathis (1985))

$$U_n \to \widetilde{U}_n = \begin{cases} U_n & n \geq 0 \\ \infty & n < 0 \end{cases} \quad . \tag{4.40}$$

This amounts to restrict the eigenvalue problem on the positive half-line. We then have to account for the boundary condition $\psi_0(z_0) = 0$, in which case the related energy eigenvalues are produced by the $w = E/\mathcal{F}$ -roots of the equation

$$\psi_0\left(z\frac{t_H}{\mathcal{F}}\right) = J_{-E/\mathcal{F}}\left(2\frac{t_H}{\mathcal{F}}\right) = 0 \quad . \tag{4.41}$$

We are ready to derive an explicit, but crude approximation to (4.41), by resorting to the leading term

$$J_\nu(z) = \sqrt{\frac{2}{\pi z}} \cos\left(z - \frac{\pi}{2}\nu - \frac{\pi}{4}\right) + O\left(\frac{1}{z}\right) \tag{4.42}$$

of the asymptotic expansion of the Bessel-function as $|z| \to \infty$. This yields an unexpected "harmonic oscillator" approximation like

$$E = E_{n_0} = \left(2n' + \frac{3}{2}\right)\mathcal{F} - \frac{4}{\pi}t_H \tag{4.43}$$

in so far as the quotient $t_H/\mathcal{F}$ exhibits sufficiently large values, where n' is an integer. Some further conversions are in order. Indeed, one has (see e.g. Abramowitz and Stegun (1972))

$$J_\nu(z_0) = \frac{(z_0/2)^\nu \exp(-iz_0)}{\Gamma(\nu+1)} M\left(\nu + \frac{1}{2}, 2\nu + 1, 2iz_0\right) \tag{4.44}$$

where

$$M(a, c, z) = \sum_{k=0}^{\infty} \frac{(a)_k}{(c)_k} \frac{z^k}{k!} \tag{4.45}$$

denotes the Kummer-function . Then (4.41) leads to the eigenvalue condition

$$\frac{1}{\Gamma(1 - E/\mathcal{F})} M\left(\frac{1}{2} - \frac{E}{\mathcal{F}}, 1 - \frac{2E}{\mathcal{F}}, 4i\frac{t_H}{\mathcal{F}}\right) = 0 \tag{4.46}$$

so that, in general, the energy-spectrum is generated by the poles of Gamma function $\Gamma(1 - E/\mathcal{F})$, as well as by the zero's of the Kummer's function. In the first case one obtains

$$1 - \frac{E}{\mathcal{F}} = 0, 1, 2, ... \tag{4.47}$$

which leads to equally spaced energy levels like

$$E = E_{n_0}^{(\Gamma)} = \mathcal{F}(1 + n_0) \tag{4.48}$$

where $n_0 = 0, 1, 2, ...$ Further eigenvalues are generated by the equation

$$M\left(\frac{1}{2} - \frac{E}{\mathcal{F}}, 1 - \frac{2E}{\mathcal{F}}, 4i\frac{t_H}{\mathcal{F}}\right) = 0 \tag{4.49}$$

but closed solutions to energies implied in this way are not available.

4.3 The Coulomb potential on the Bethe-lattice

An interesting solvable discrete model is the Coulomb-potential on the Bethe-lattice (Gallinar (1984)). This lattice is characterized by a central point x_0 surrounded in a hierarchical manner by sets of points like $x^{(c)}_{j,l_j}$, where $j = 1, 2, 3, \ldots$. This latter number specifies the increasing complexity of discrete structures, such that

$$l_j = 1, 2, \ldots, c\,(c-1)^{j-1} \quad . \tag{4.50}$$

The corresponding coordination parameter is denoted by $c \geq 2$. This means that any point of this lattice is surrounded by c nearest neighbors. The eigenvalue equation is now given by the discrete Schrödinger-equation

$$\mathcal{H}\psi_{j,l_j} = t_H \psi_{j-1,l_{j-1}} + t_H \psi_{j-1,l_{j+1}} + V_j \psi_{j,l_j} = E' \psi_{j,l_j} \tag{4.51}$$

which relies on (4.3), where

$$V_j = \frac{\alpha_0}{j} \tag{4.52}$$

and $\alpha_0 > 0$. We shall also consider that $|V_0| = \infty$, which makes the problem well-defined on the positive discrete half-line, as shown in the previous section. Equation (4.51) has also been solved by applying the Green-function method, such as discussed before in terms of corresponding tight binding Hamiltonians (Brinkmann and Rice (1970), Hubbard (1979)). This Green-function exhibits the form

$$G_j^{(-)}\left(E' - i\varepsilon\right) = \langle j, l_j| \, \frac{1}{E' - i\varepsilon - \mathcal{H}} \, |j, l_j\rangle \quad . \tag{4.53}$$

Then the energy eigenvalues rely, in general, both on the poles of the real part

$$ReG_j^{(-)}\left(E' - i\varepsilon\right) = \langle j, l_j| \, P\frac{1}{E' - \mathcal{H}\varkappa} \, |j, l_j\rangle \tag{4.54}$$

as well as on the discrete scars of the Dirac delta function characterizing the imaginary part of (4.53), i.e.

$$D_j\left(E'\right) = \frac{1}{\pi} ImG_j^{(-)}\left(E' - i\varepsilon\right) = \langle j, l_j| \, \delta\left(E' - \mathcal{H}\right) |j, l_j\rangle \quad . \tag{4.55}$$

This equation leads to the density of states per unit volume (see also sections 6.2 and 8.4) by performing, of course, the sum over states. The interesting point is that (4.53) can be handled in terms of the incomplete beta-function, which results in an explicit derivation of the discrete singular part of the density of states. Such issues lead to the energy eigenvalue

$$E' \equiv -E = \sqrt{4\,(c-1)\,t_H^2 + \frac{\alpha_0^2}{n_0^2}} \tag{4.56}$$

exhibiting a rather special square-root behavior, where $n_0 = j = 1, 2, 3, \ldots$. Note that such results have been obtained without resorting to the eigenfunction. One sees that

$$E + 2 \rightarrow E_S = -\frac{\alpha_0^2}{4n_0^2} \tag{4.57}$$

as $n_0 \rightarrow \infty$, which reproduces the discrete spectrum of the usual hydrogen atom via $t_H = 1$ and $c = 2$. In this latter case, the Bethe-lattice gets identified with the discrete positive half-line.

4.4 The discrete s-wave description of the Coulomb-problem

An alternative discrete analog of the s-wave ($l = 0$) Coulomb-problem looking like

$$\Delta\nabla\psi\,(x) + \left(E + 2 + \frac{\alpha_0}{x}\right)\psi\,(x) = 0 \tag{4.58}$$

has also been proposed by Kvitsinsky (1992). This works in accord with (4.1), where $x = 0, 1, 2, \ldots$ denotes the dimensionless discrete radial coordinate. Accordingly $r = xa_0$ and $\alpha_0 = e^2/a_0$, where a_0 denotes a suitable length scale. One could eventually try to solve (4.58) in terms of (1.11), i.e. with the help of hypergeometric type solutions. However, this fails to work as one would have $\tau(x) = 0$, in which case $\alpha_0 = \lambda = 0$ by virtue of (1.77). We shall then start from the wavefunction-ansatz

$$\psi\,(x) = x\exp\,(i\theta x)\,L\,(x) \tag{4.59}$$

working in combination with the energy-dispersion law

$$E+2=E(k)=2(1-\cos\theta) \tag{4.60}$$

where $\theta = ka_0$, which is reminiscent to (4.104) in Freeman et al (1969). Inserting (4.59) into (4.58) yields the discrete equation

$$\begin{aligned}\exp(i\theta)(x+1)L(x+1)+\exp(-i\theta)(x-1)L(x-1)+ \\ +(\alpha_0-2x\cos\theta)L(x)=0\end{aligned} \tag{4.61}$$

which can be rewritten equivalently as

$$\begin{aligned}(x+1)L(x+1)+\exp(-2i\theta)(x-1)L(x-1)+ \\ +\exp(-i\theta)(\alpha_0-2x\cos\theta)L(x)=0 \quad .\end{aligned} \tag{4.62}$$

On the other hand, the hypergeometric function

$$F(a)={}_2F_1(a,b,c;\xi) \tag{4.63}$$

obeys the discrete equation

$$a(\xi-1)F(a+1)+(c-a)F(a-1)+[2a-c-(a-b)\xi]F(a)=0 \tag{4.64}$$

in accord with (1.115), i.e. by virtue of Gauss's relations for contiguous hypergeometric functions. This relies on (4.62) via $L(x)=F(a)$ and

$$a=-x+\beta \tag{4.65}$$

where β remains to be established.

Comparing (4.64) and (4.62) yields the matching conditions

$$a(\xi-1)=(x-1)\exp(-2i\theta)=(\beta-(a+1))\exp(-2i\theta) \tag{4.66}$$

$$c-a=\beta-a+1 \tag{4.67}$$

and

$$2a-c-(a-b)\xi=\exp(-i\theta)(\alpha_0-2(\beta-a)\cos\theta) \quad . \tag{4.68}$$

Equation (4.66) gives

$$a\left(\xi - 1 + \exp\left(-2i\theta\right)\right) = \exp\left(-2ik\right)\left(\beta - 1\right) \tag{4.69}$$

which can be readily satisfied in terms of the inter-connected relationships

$$\xi = 1 - \exp\left(-2i\theta\right) \tag{4.70}$$

and

$$\beta = 1 \tag{4.71}$$

in which case

$$c = 2 \quad . \tag{4.72}$$

On the other hand, one obtains

$$b = i\tan\theta\left(1 - \frac{i\alpha_0}{2\sin\theta}\right) \tag{4.73}$$

by virtue of (4.68).

So we found that (4.58) exhibits the solution

$$\psi\left(x\right) = x\exp\left(i\theta\right)_2 F_1\left(-x+1, b, 2; 1 - \exp\left(-2i\theta\right)\right) \tag{4.74}$$

which is valid irrespective of E. Looking however, for bound-states we have to realize that the square integrability of $\psi\left(x\right)$ is able to be fulfilled if

$$\theta = \theta_n = i\mathcal{K}_n \tag{4.75}$$

in which $\mathcal{K}_n > 0$, provided that the hypergeometric function becomes a polynomial of degree n_r, where $n_r = 0, 1, 2, \ldots$ plays again the role of the radial quantum number. In this latter case, one should have $b = -n_r$ in accord with (1.37), so that

$$u - \frac{\alpha_0}{2} = n_r\sqrt{1+u^2} \tag{4.76}$$

where

$$u = \sinh\left(\mathcal{K}_n\right) \quad . \tag{4.77}$$

There is also an alternative to (4.76), namely $a = 1 - x = -n_r$, but this has a trivial meaning. The appropriate solution is then given by

$$\sinh\left(\mathcal{K}_n\right) = \frac{\alpha_0}{2}\frac{1}{n_r + 1} \tag{4.78}$$

which produces the energy of the discrete s-wave Coulomb-system as

$$E = E_{n_r} = -2\left(1 + \frac{\alpha_0^2}{4\left(n_r + 1\right)^2}\right)^{1/2} \tag{4.79}$$

in accord with (4.60) and (4.75). This result is similar to (4.56) derived before. The continuous limit is then given by

$$E + 2 \to E_S = -\frac{\alpha_0^2}{4\left(n_r + 1\right)^2} \tag{4.80}$$

which reproduces precisely the usual s-wave result.

One sees that (4.79) and (4.80) proceed in a close analogy with (4.56) and (4.57). We have to realize that both (4.56) and (4.79) are interesting, as they incorporate square-root energy contributions, which are almost typical for relativistic energy descriptions. This latter interpretation is confirmed by the energy-formula established within the relativistic quasipotential approach to the Coulomb-problem. Indeed, (4.107) reproduces identically (4.79) as soon as $l = 0$, which represents a rather remarkable agreement.

4.5 The Maryland class of potentials

The potential

$$V_M\left(x\right) = \frac{a + bz}{1 + cz} \tag{4.81}$$

where $z = q^x$ has also been analyzed in some more detail (Kvitsinsky (1994)). Here $q \in \mathbb{C}$ stands for an inherent deformation parameter producing the conversion of the second-order discrete equation like (4.3) into a q-difference one. This potential is of special interest, as it is able to reproduce well-known potentials, such as the Maryland-model (Grempel et al (1982), Bentosela et al (2003)), the exponential and Hulthén potentials, for selected values of parameters a, b and c, as shown in Table 4.1. One has

$q = \exp(-2\pi i\alpha)$ for the Maryland-model , whereas the $q = \exp(-\alpha) > 0$ -in the case of exponential and Hulthén-potentials.

Table 4.1. Concrete realizations of the Maryland potential for selected values of the parameters a, $b$and c.

q	a	b	c	$V(x)$
$\exp(-2\pi i\alpha)$	$-i\lambda$	$i\lambda\exp(-2i\theta)$	$\exp(-2i\theta)$	$\lambda\tan(\pi\alpha x+\theta)$
$\exp(-\alpha)$	0	λ	0	$\lambda\exp(-\alpha x)$
$\exp(-\alpha)$	0	λ	-1	$\lambda\exp(-\alpha x)/(1-\exp(-\alpha x))$

One remarks that the wavefunction can be rewritten as

$$\psi(x) = \varphi(q^x) \tag{4.82}$$

in which case (4.3) exhibits the q-difference form

$$\varphi(qz) + \varphi(z/q) + \frac{a+bz}{1+cz}\varphi(z) = E'_M\varphi(z) \equiv E\varphi(z) \tag{4.83}$$

referred to above. The identification $E'_M = E$ is used just in connection with (4.3). Multiplying (4.83) by $1+cz$, opens the way to use the power-series expansion

$$\varphi(z) = z^\sigma \sum_{n=0}^{\infty} f_n z^n \tag{4.84}$$

where $f_0 \neq 0$ and where σ is a positive parameter. This results in the two-term recurrence relation

$$\left(b - cE + cq^{n+\sigma-1} + \frac{c}{q^{n+\sigma-1}}\right) f_{n-1} = \left(E - a - q^{n+\sigma} - \frac{1}{q^{n+\sigma}}\right) f_n \tag{4.85}$$

where $n = 0, 1, 2, \ldots$. Inserting $n = 0$, yields the energy-dependent relationship

$$E - a = q^\sigma + \frac{1}{q^\sigma} \tag{4.86}$$

which works in accord with the condition $f_{-1} = 0$. Furthermore, one finds

$$f_{n+1} = \frac{\prod\limits_{k=0}^{n} \left[(b - cE) + c\left(q^{k+\sigma} + 1/q^{k+\sigma}\right)\right]}{\prod\limits_{k=0}^{n} \left[(q^{\sigma} + 1/q^{\sigma}) - (q^{\sigma+k+1} + 1/q^{\sigma+k+1})\right]} f_0 \quad . \tag{4.87}$$

This exhibits a polynomial solution, say

$$\varphi(z) = P_{n_0}(z) \tag{4.88}$$

if

$$b - cE + c\left(q^{n_0+\sigma} + \frac{1}{q^{n_0+\sigma}}\right) = 0 \quad . \tag{4.89}$$

The degree of this polynomial is denoted by n_0. In this case, the energy comes from the algebraic equation

$$q^{\pm n_0} = \frac{E - \frac{b}{c} \pm \sqrt{\left(E - \frac{b}{c}\right)^2 - 4}}{E - a \pm \sqrt{(E-a)^2 - 4}} \tag{4.90}$$

provided that $a \neq b/c$, which also means that it is enough to restrict ourselves to the "+" sign. The q-dependence of the energy in (4.90) is displayed in figure (4.1), for $0 < q < 1$, $n_0 = 3$ (left curve) and $n_0 = 1$ (right curve). We have assumed, for instance, that $a = 2$, and $b/c = 1$. Both energies are negative and increase with q up to the upper bound $E = -2$. We can also say that underlying q-values move to the right, when n_0 becomes larger.

Assuming that n_0 is not an integer, we have to resort, however, to non-polynomial solutions. This leads us to rewrite (4.89) as

$$b - cE + c\left(q^{r} + \frac{1}{q^{r}}\right) = 0 \quad . \tag{4.91}$$

This indicates that polynomial (non-polynomial) solutions are involved if $r = n_0 + \sigma$ ($r \neq n_0 + \sigma$). A such discrimination looks reasonable, as the difference $r - \sigma$ may be actually an integer, or not. In the latter case (4.85) yields

$$\frac{f_{n+1}}{f_n} = -c\frac{q^{r} + 1/q^{r} - q^{n+\sigma} - 1/q^{n+\sigma}}{q^{\sigma} + 1/q^{\sigma} - q^{n+\sigma+1} - 1/q^{n+\sigma+1}} \tag{4.92}$$

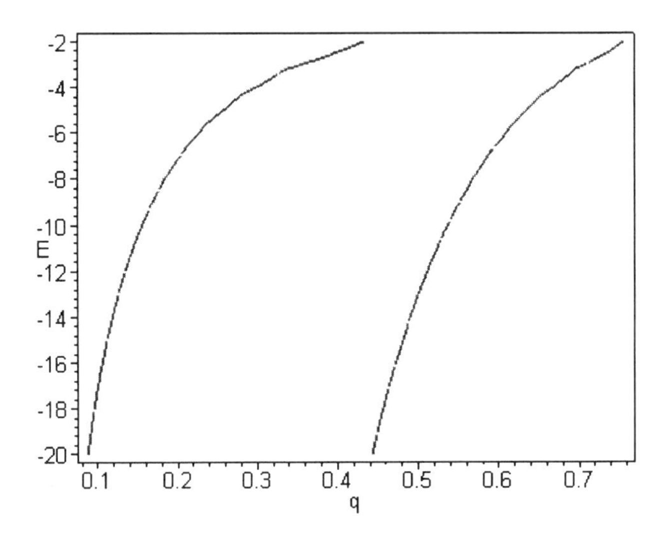

Fig. 4.1 **The q-dependence of the energy for $b/c = 1$, $a = 2$, $n_0 = 3$ (right curve) and $n_0 = 1$ (left curve).**

which relies on a hypergeometric function of the ${}_2F_1$-kind. Indeed, comparing (1.46) and (4.92) yields the identifications $p_1 = 2$, $p_2 = 1$, $\gamma = -cq$, $\beta_1 = 2\sigma + 1$, $\alpha_1 = \sigma - r$ and $\alpha_2 = \sigma + r$. So, the non-polynomial solution reads

$$\psi(x) = {}_2F_1^{(q)} \left(\begin{array}{c} \sigma - r,\ \sigma + r \\ \\ 2\sigma + 1 \end{array} \middle| -cq^{x+1} \right) q^{\sigma x} \tag{4.93}$$

so that

$$\psi(0) = {}_2F_1^{(q)} \left(\begin{array}{c} \sigma - r,\ \sigma + r \\ \\ 2\sigma + 1 \end{array} \middle| -cq \right) . \tag{4.94}$$

Note that a boundary condition like

$$\psi(0) = 0 \tag{4.95}$$

which is reminiscent to (4.40), has also been proposed in order to derive the energies characterizing even-parity states (Kvitsinsky (1994)).

4.6 The relativistic quasipotential approach to the Coulomb-problem

Besides Snyder's space time discretization referred to before, further interplays between relativity and space discreteness are provided by the relativistic quasipotential description. Indeed, handling the relativistic two-body problem on the Lobachevsky-space, proofs have been given that a relativistic configuration space, say $\overrightarrow{R} = \overrightarrow{n}\mathcal{R}$, can be introduced in a suitable manner (Kadyshevsky et al (1968)). For this purpose the usual Fourier decompositions in terms of plane waves are replaced by novel decompositions in terms of relativistic "plane waves" like

$$\zeta\left(\overrightarrow{p}, \overrightarrow{R}\right) = \left(\frac{p^{\mu} R_{\mu}}{m_0 \mathcal{R}}\right)^{-1-(i/\hbar)m_0\mathcal{R}} \tag{4.96}$$

which corresponds to unitary representations of the Lorentz-group. There is $R^{\mu} = \left(\mathcal{R}, \overrightarrow{R}\right)$ and $p^{\mu} = (p_0, \overrightarrow{p})$, such that $R^{\mu}p_{\mu} = p_0\mathcal{R} - \overrightarrow{p} \cdot \overrightarrow{R}$ and $p_o = (\overrightarrow{p}^2 + m_0^2)^{1/2}$. This means that $\mathcal{R}$ is the eigenvalue of a relativistic radial coordinate operator such as

$$R_{op} = i\frac{\hbar}{m_0}\left[p^{\mu}\frac{\partial}{\partial p^{\mu}} + 1\right] \tag{4.97}$$

in which case

$$R_{op}\zeta\left(\overrightarrow{p}, \overrightarrow{R}\right) = \mathcal{R}\zeta\left(\overrightarrow{p}, \overrightarrow{R}\right) \quad . \tag{4.98}$$

In this context the relativistic free particle Hamiltonian is given by the finite difference operator

$$H_0 = 2\cosh\left(i\frac{\partial}{\partial r}\right) + \frac{2i}{r}\sinh\left(i\frac{\partial}{\partial r}\right) - \frac{\Delta_{\vartheta,\varphi}}{r^2}\exp\left(i\frac{\partial}{\partial r}\right) \tag{4.99}$$

where $\Delta_{\vartheta,\varphi}$ denotes the usual angular Laplacian, such that the eigenvalue problem reads

$$H_0\zeta\left(\overrightarrow{p}, \overrightarrow{r}\right) = 2E_p\zeta\left(\overrightarrow{p}, \overrightarrow{r}\right) \quad . \tag{4.100}$$

There is $E_p = p_0$, while

$$r = \frac{m_0 c}{\hbar}R \tag{4.101}$$

denotes the relativistic dimensionless coordinate.

The relativistic quasipotential description with a local potential is then given by the difference equation (Kadyshevsky et al (1968), Freeman et al (1969))

$$\left(H_0^{(r)} + V(r)\right) \psi_p^{(l)}(r) = 2E_p \psi_p^{(l)}(r) \tag{4.102}$$

where

$$H_0^{(r)} = 2\cosh\left(i\frac{\partial}{\partial r}\right) + \frac{2i}{r}\sinh\left(i\frac{\partial}{\partial r}\right) + \frac{l(l+1)}{r^2}\exp\left(i\frac{\partial}{\partial r}\right) \tag{4.103}$$

plays the role of the radial Hamiltonian, in which in general, $V(r) = V(r; E_p)$. For convenience, the spherical symmetry has been assumed. The above equation, though being relativistic, is similar to a non-relativistic Schrödinger-equation. In particular, the Coulomb-potential $V(r) = -\alpha_0/r$ has been discussed by using the parametrization

$$E_p = \cos\chi_p \quad . \tag{4.104}$$

Inserting (4.104) in (4.102), it has been found that the Coulomb-problem is solvable in terms of the hypergeometric function ${}_2F_1(a, b, c; z)$,. The solution is then given by (Freeman et al (1969))

$$\psi_p^{(l)}(r) = C(l, \chi_p)\exp(-r\chi_p)(-r)^{l+1} F(a, b, c; z) \tag{4.105}$$

which proceeds up to a normalization factor, where

$$z = 2i\sin\chi_p \exp(-\chi_p) \tag{4.106}$$

$$a = l + 1 - \frac{\alpha_0}{2\sinh\chi_p} \tag{4.107}$$

$$b = l + 1 - ir \tag{4.108}$$

and

$$c = 2l + 2 \quad . \tag{4.109}$$

Such equations can be viewed, in several respects, as a $l \neq 0$, generalization of the ones presented in section 4.4. The Coulomb energy is then given by the polynomial reduction

$$\sinh \chi_p = \frac{\alpha_0}{2n} \tag{4.110}$$

which corresponds to (4.78), such that $n = 1, 2, \ldots$.

Furthermore, we have to consider that $a = -n_r$, in which case the principal quantum number is given by

$$n = n_r + l + 1 \tag{4.111}$$

as usual. This leads to the energy

$$2E_p = 2E_p^{(n)} = -2\left(1 + \frac{\alpha_0^2}{4n^2}\right)^{1/2} \tag{4.112}$$

which reproduces (4.79) just by inserting $l = 0$. This agreement confirms, once more again, our interpretation that the presence of the square-root in the expression of the energy can be viewed as a well defined signature of a relativistic background.

4.7 The infinite square well

Our last example in this chapter is the infinite square well proceeding on a lattice with N_s sites. Using e.g. (4.3) yields the discrete equation

$$\psi_{n+1} + \psi_{n-1} = E'\psi_n \tag{4.113}$$

where $n \in (0, N_s)$, such that $V_n = -\infty$ for $n \notin (0, N_s)$. This produces the zero boundary conditions

$$\psi_{n=0} = \psi_{n=N_s} = 0 \tag{4.114}$$

which have to be handled in terms of the ansatz

$$\psi_n = C \sin(np) \quad . \tag{4.115}$$

The energy is then given by

$$E' = 2\cos p = 2\cos\left(\frac{k\pi}{N_s}\right) \tag{4.116}$$

where k is a positive integer, which corresponds to (11) in Boykin and Klimeck (2004). The normalization

$$\sum_{n=0}^{N_s-1} |\psi_n|^2 = 1 \tag{4.117}$$

can also be invoked, in which case

$$\psi_n = \psi_n^{(k)} = \sqrt{\frac{2}{N_s}} \sin\left(n\frac{k\pi}{N_s}\right) \quad . \tag{4.118}$$

The present energy exhibits the form

$$E_1 = E_{SW}^{(k)} = 2 - E' = 4\sin^2\left(n\frac{k\pi}{N_s}\right) \tag{4.119}$$

in accord with (4.1), so that $E_1 \in (0, 4]$.

4.8 Other discrete systems

Other discrete systems such as produced e.g. by the selected wavefunction

$$\varphi_q(n, \lambda) = \frac{\left[\lambda + \left(\lambda^2 - q\right)^{1/2}\right]^n - \left[\lambda - \left(\lambda^2 - q\right)^{1/2}\right]^n}{2\left(\lambda^2 - q\right)^{1/2}} \tag{4.120}$$

which exhibits the limit

$$\varphi_q(n, \lambda) \to (2\lambda)^{n-1} \tag{4.121}$$

as $q \to 0$, can also be shortly mentioned. This satisfies the discrete equation

$$\varphi_q(n+1, \lambda) + q\varphi_q(n-1, \lambda) = 2\lambda\varphi_q(n, \lambda) \tag{4.122}$$

which has been studied before in connection with the inverse scattering method (Karlo et al (1995)). This latter equation generalizes (4.113) towards $q \neq 1$. Proofs have also been given that the discrete versions of the

Korteweg- de Vries, sine-Gordon and Liouville equations are able to rely on the discrete Schrödinger equation

$$\psi(n+2) + V(n)\psi(n+1) = \lambda\psi(n) \tag{4.123}$$

where $V(n)$ plays the role of the potential (Boiti et al (2003)). One sees that (B.14) reproduces the above equation via $a_{n+1} = -1$, $d_{n+1} = \lambda$ and

$$V(n) = E + v_{n+1} \quad . \tag{4.124}$$

A similar inter-connection can also be established with (5.119).

Chapter 5

Discrete Analogs and Lie-Algebraic Discretizations. Realizations of Heisenberg-Weyl Algebras

Now we look for a more general and deeper understanding of tight binding equations from the mathematical point of view. So far the discrete Schrödinger-equations discussed in the previous chapter are expressed by typical superpositions between the kinetic-energy and the potential, as usual. However, there are discrete formulations working in terms of orthogonal polynomials of the discrete variable which go beyond transparent superpositions referred to above. In this latter context one deals with discrete analogs of quantum-mechanical systems. We shall then discuss discrete analogs of the harmonic oscillator (Lorente (2001b), Atakishiyev and Suslov (1990)) and of the Coulomb-system (Lorente (2001b). The study of algebraic symmetries plays also an important role as it serves for explicit investigations of spectral structures, too. However, there are other mathematical developments deserving further attention. This concerns discrete realizations of the Heisenberg-Weyl commutation relation (Zhedanov (1993)) $HR = qRH$, where e.g. $0 < q < 1$, which arises in the theory of the Heisenberg-Weyl group (Boon (1972)), or in the description of the q-deformed algebras (Faddeev et al (1990), Majid (1990), Manin (1988)). It should also be mentioned that difference operator realizations of both $SL(N)$ and $SL_q(N)$ groups have been discussed in some more detail before (Miller (1969), Miller (1970), Floreanini and Vinet (1995a), Floreanini and Vinet (1995b), Shafiekhani (1994), Vilenkin and Klimyk (1992)).

On the other hand studies in the general theory of integrable systems have been done by resorting to the Bethe-ansatz method (Gaudin (1983), Faddeev et al (1995b), Korepin et al (1992)). Concrete quantum-mechanical manifestations of this latter method will then be presented in Appendix B.

5.1 Lie algebraic approach to the discretization of differential equations

A theoretical understanding of difference equations, their solvability included, is able to be expressed, interestingly enough, in terms of algebraic attributes of $\delta_{\pm}$ operators introduced in section 1.1. Indeed, the difference operators defined in accord with (1.1) and (1.2) are able to produce a realization of the Heisenberg algebra

$$[A, B] = 1 \tag{5.1}$$

via

$$A = \delta_{+} \tag{5.2}$$

and

$$B = x\left(1 - a\delta_{-}\right) \tag{5.3}$$

in which case

$$ABf\left(x\right) = f\left(x\right) + x\delta_{-}f\left(x\right) \tag{5.4}$$

and

$$BAf\left(x\right) = x\delta_{-}f\left(x\right) \quad . \tag{5.5}$$

The algebraic attributes referred to above are provided by the action of difference operators on quasi-monomials like (Smirnov and Turbiner (1995))

$$x^{(n+1)} = x\left(x-a\right)\left(x-2a\right)\cdots\left(x-na\right) = \delta^{n+1}\frac{\Gamma\left(x/a+1\right)}{\Gamma\left(x/a-n\right)} \tag{5.6}$$

where $x^{(1)} = x$, so that $x^{(n+1)} \to x^{n+1}$ if $a \to 0$. So, it can be easily verified that

$$Ax^{(n)} = nx^{(n-1)} \tag{5.7}$$

$$Bx^{(n)} = x^{(n+1)} \ , \tag{5.8}$$

as well as (5.1), can be viewed as well-defined difference counterparts of the continuous relationships

$$\frac{d}{dx}x^n = nx^{n-1} \tag{5.9}$$

$$x\ x^n = x^{n+1} \tag{5.10}$$

and

$$\left[\frac{d}{dx}, x\right] = 1 \tag{5.11}$$

respectively.

Under such circumstances we are ready to say that the difference counterpart of the second-order linear differential operator $L\left[d/dx, x\right]$, for which

$$L\left[\frac{d}{dx}, x\right]\varphi(x) = E\varphi(x) \tag{5.12}$$

where L is synonymous with the Hamiltonian given by

$$L\left[A, B\right]\varphi(B)|0\rangle = E\varphi(B)|0\rangle \quad . \tag{5.13}$$

The vacuum state is defined via $A|0\rangle = 0$, as usual. Using in an explicit manner the x-representation, yields

$$\langle x|\, A\,|0\rangle \equiv A_x f(x) = 0 \tag{5.14}$$

where $\langle x|\ 0\rangle = f(x)$ and $A \equiv A_x$. This means that

$$f(x+a) = f(x) \tag{5.15}$$

in accord with (5.2). Leaving aside the study of such periodic functions, we shall make the simplifying identification $f(x) = 1$. Then the function $B^n|0\rangle$ can be handled as follows

$$\langle x|\, B^n\,|0\rangle = B_x^n f(x) = x^{(n)} \tag{5.16}$$

which indicates that the continuous polynomial

$$\varphi(x) = \sum c_k x^k \tag{5.17}$$

is replaced by

$$\widetilde{\varphi}(x) = \sum c_k x^{(k)} \quad . \tag{5.18}$$

Accordingly, the difference analog of (5.12) is

$$L[A, B]\,\widetilde{\varphi}(x) = E\widetilde{\varphi}(x) \tag{5.19}$$

which shows that the original eigenvalue is preserved. In other words, the Lie algebraic approach to the discretization done above exhibits the isospectral property, too.

5.2 Describing exactly and quasi-exactly solvable systems

It is well known that the dynamical symmetry characterizing exactly solvable systems is done by the $SL(2)$-group (Debergh et al (2002), Turbiner (1992)). In the present case, the generators of this algebra are given by

$$J_n^+ = B^2 A - nB \tag{5.20}$$

$$J_n^- = A \tag{5.21}$$

and

$$J_n^0 = BA - \frac{n}{2} \tag{5.22}$$

where n denotes a non-negative integer. Thus, one has

$$\left[J_n^0, J_n^\pm\right] = \pm J_n^\pm \tag{5.23}$$

and

$$\left[J_n^+, J_n^-\right] = -2J_n^0 \quad . \tag{5.24}$$

The Hamiltonian characterizing exactly solvable systems is given by a cubic polynomial in the sole generators $J^0 \equiv J_0^0$ and $J^- \equiv J_0^-$ only. This polynomial reads

$$L_{ES}\left[B,A\right]=A_1J^0J^0\left(J^-+\frac{1}{\delta}\right)+A_2J^0J^-+ \qquad (5.25)$$

$$+A_3J^0+A_4J^-+A_5$$

where A_i $(i=1,2,\ldots,5)$ are suitable coefficients. Accordingly, one obtains the second-order difference equation (Smirnov and Turbiner (1995))

$$\left(\frac{A_4}{\delta}+\frac{A_2}{\delta^2}+\frac{A_1}{\delta^3}+x^2\right)\psi\left(x+\delta\right)+ \qquad (5.26)$$

$$+\left(A_5-\frac{A}{\delta}+\left(\frac{A_1}{\delta^2}-2\frac{A_2}{\delta^2}+\frac{A_3}{\delta}\right)x-2\frac{A_1}{\delta^3}x^2\right)\psi\left(x\right)+$$

$$+\left(-\left(\frac{A_1}{\delta^2}-\frac{A_2}{\delta^2}+\frac{A_3}{\delta}\right)x+\frac{A_1}{\delta^3}x^2\right)\psi\left(x-\delta\right)=E\psi\left(x\right)$$

where

$$E=E_n=\frac{1}{\delta}A_1n^2+A_3n \qquad (5.27)$$

and $n=0,1,2,\ldots$. One sees that the underlying Hamiltonian is of a hypergeometric type (Nikiforov, Suslov and Uvarov (1991), Smirnov and Turbiner (1995))

$$L_{ES}\left[\frac{d}{dx},x\right]=Q_2\left(x\right)\frac{d^2}{dx^2}+Q_1\left(x\right)\frac{d}{dx}+Q_0 \qquad (5.28)$$

where the $Q_k\left(x\right)$ polynomials $(k=1,2,3)$ are of the kth -order. This corresponds to (1.11), so that mutual identifications of coefficients can be easily done.

In the case of quasi-exactly solvable systems only a finite amount of polynomial eigenfunctions are able to be established (Shifman (1989a), Turbiner and Ushveridze (1987), Ushveridze (1994)). In this case, the Hamiltonian is expressed by the n-dependent superposition

$$L_{QES}\left[B,A\right]=A_6\left(J_n^++\delta J_n^0J_n^0\right)+A_1J_n^0J_n^0\left(J_n^-+\frac{1}{\delta}\right)+ \qquad (5.29)$$

$$+A_2J_n^0J_n^-+A_3J_n^0+A_4J_n^-+A_5$$

where the A_i's $(i=1,2,\ldots,6)$ are new pertinent coefficients. Further developments of such issues, like quantum-group $SL_q(2)$ -symmetries will be discussed in sections 9.4 and 10.2. However, we have to say that there are quasi-exactly solvable Hamiltonians which do not exhibit the $SL(2)$ -symmetry displayed above (Jatkar et al (1989)).

5.3 The discrete analog of the harmonic oscillator

A difference analog of the harmonic oscillator has been discussed in terms of Krawtchouk polynomials by Atakishiyev and Suslov (1990), Atakishiyev and Wolf (1997) as well as by Lorente (2001b). The Hamiltonian is given by

$$\mathcal{H} = -\frac{1}{2}\left[\alpha(x)\exp\left(-\frac{\partial}{\partial x}\right) + \alpha(-x)\exp\left(\frac{\partial}{\partial x}\right)\right] + \frac{1}{2}(N_0+1) \quad , \tag{5.30}$$

so that

$$\mathcal{H}\psi(x) = -\frac{1}{2}\alpha(x)\psi(x-1) - \frac{1}{2}\alpha(-x)\psi(x+1) + \frac{1}{2}(N_0+1)\psi(x) = \tag{5.31}$$

$$= E\psi(x)$$

where

$$\alpha(x) = \sqrt{\left(\frac{N_0}{2}+x\right)\left(\frac{N_0}{2}-x+1\right)} \quad . \tag{5.32}$$

Introducing a further operator like

$$J_2 = -\frac{i}{2}\left[\alpha(x)\exp\left(-\frac{\partial}{\partial x}\right) - \alpha(-x)\exp\left(\frac{\partial}{\partial x}\right)\right] \tag{5.33}$$

one remarks that one obtains a realization of the $SO(3)$-group

$$[J_j, J_k] = \varepsilon_{jkl}J_l \tag{5.34}$$

in terms of the identifications

$$J_1 = x \tag{5.35}$$

and

$$J_3 = \mathcal{H} - \frac{1}{2}(N_0+1) \tag{5.36}$$

where the l-summation is implicitly understood.. The Casimir-operator can also be easily calculated. One has

$$\mathbb{C}_0 = J_1^2 + J_2^2 + J_3^2 \equiv l\,(l+1) \tag{5.37}$$

in which $l = N_0/2$. This shows that the irreducible representations are characterized by the $N_0 + 1$-dimension.

Next, we have to invoke the standard eigenvalue equation

$$J_3\,|l,m\rangle = m\,|l,m\rangle \tag{5.38}$$

where $m = -l, -l+1, \ldots, l$. This shows that oscillator energies are produced by

$$\mathcal{H}\left|\frac{N_0}{2}, m\right\rangle = \left(n + \frac{1}{2}\right)\left|\frac{N_0}{2}, m\right\rangle \tag{5.39}$$

where now

$$n = \frac{N_0}{2} + m = 0, 1, \ldots, N_0 \quad . \tag{5.40}$$

One sees, of course, that now n is bounded from above. A such limitation can be viewed as a typical manifestation of space-discreteness (see also Kehagias and Zoupanos (1994)).

The corresponding eigenfunction can be established in terms of normalized Krawtchouk-functions (Lorente (2001a)):

$$K_n^{(p)}\,(x, N_0) = \sqrt{\frac{n!\,(N_0-n)!}{(pq)^n}\,\frac{p^x q^{N_0-x}}{x!\,(N_0-x)!}}\,k_n^{(p)}\,(x, N_0) \tag{5.41}$$

where $p+q = 1$ ($p > 0$ and $q > 0$) and where $k_n^{(p)}\,(x, N_0)$ is the Krawtchouk-polynomial (Nikiforov, Suslov and Uvarov (1991), Atakishiyev and Suslov (1990), Lorente (2001b)), for which main formulae have been presented in the Appendix A. This function is normalized as

$$\sum_{x=0}^{N_0+1} K_n^{(p)}\,(x, N_0)\,K_{n'}^{(p)}\,(x, N_0) = \delta_{n',n} \quad . \tag{5.42}$$

Indeed, the shifted Krawtchouk-function

$$\psi_n\,(x) = K_n^{(p)}\left(n + \frac{N_0}{2}, N_0\right) \tag{5.43}$$

obeys the equation (see also equation (NK1) in Lorente (2001a))

$$-\frac{1}{2}\alpha(x)\psi_n(x-1)-\frac{1}{2}\alpha(-x)\psi_n(x+1)=m\psi_n(x) \tag{5.44}$$

if $p=q=1/2$, which reproduces identically (5.31) in terms of (5.40). For the sake of illustration some concrete expressions of shifted Krawtchouk polynomials :

$$\widetilde{k}_n \equiv \widetilde{k}_n^{(1/2)}(x,N_0)=k_n^{(1/2)}\left(x+\frac{N_0}{2},N_0\right) \tag{5.45}$$

are presented herewith. So, one has $\widetilde{k}_0 = 1$, $\widetilde{k}_1 = x$ and $\widetilde{k}_2 = \left(x^2-N_0/4\right)/2$, whereas (Atakishiyev and Wolf (1997))

$$\widetilde{k}_3=\frac{1}{3!}[x^3-\frac{1}{4}(3N_0-2)x] \tag{5.46}$$

$$\widetilde{k}_4=\frac{1}{4!}[x^4-\frac{1}{2}(3N_0-4)x^2+\frac{3}{16}N_0(N_0-2)] \tag{5.47}$$

and

$$\widetilde{k}_5=\frac{1}{5!}[x^5-\frac{5}{2}(N_0-2)x^3+\frac{1}{16}\left(15N_0^2-50N_0+24\right)x] \quad . \tag{5.48}$$

For other details the original paper by Krawtchouk (1929) can be recommended.

The classical limit of (5.31) can also be readily done with the help of the $N_0 \to \infty$-limit. One finds

$$\mathcal{H}\psi(x) \to \frac{N_0+1}{2}\left(-\frac{1}{2}\frac{d^2\psi(x)}{dx^2}\right)+\frac{2x^2}{N_0}\frac{1}{2}\psi(x) \tag{5.49}$$

which shows that the x-coordinate can be rescaled as

$$x=\gamma\zeta=\left(\frac{N_0(N_0+1)}{4}\right)^{1/4}\zeta \cong \sqrt{\frac{N_0}{2}}\zeta \quad . \tag{5.50}$$

So, the leading classical equation characterizing this $N_0 \to \infty$-limit is

$$-\frac{1}{2}\frac{d^2\widetilde{\psi}(\zeta)}{d\zeta^2}+\frac{1}{2}\zeta^2\widetilde{\psi}(\zeta)=\left(n+\frac{1}{2}\right)\widetilde{\psi}(\zeta) \tag{5.51}$$

but now n ceases to be bounded from above ($n = 0, 1, 2, \ldots$). It is clear that $\widetilde{\psi}(\zeta)$ stands for the classical limit of $\psi(x)$. This reproduces precisely the eigenvalue equation of the one dimensional $1D$ harmonic oscillator , this time in terms of the ζ-coordinate needed. The corresponding eigenfunctions are the well known Hermite-Gauss functions (Flügge (1971)).

We can then conclude that the Hamiltonian (5.30) plays actually the role of a discrete analog of the harmonic oscillator . The interesting point is that the harmonic potential is included in an inherent manner into the kinetic energy. This suggests that the harmonic oscillator on the discrete space exhibits some of the attributes characterizing a "pure" gauge field. In addition, we have to realize that a difference equation is a quite complex entity, as it is able to produce, besides (5.51), an infinite set of higher-derivative equations. However, there are reasons to say these latter equations have to be viewed as being irrelevant. We have also to remark that the limitation concerning the number of admissible levels such as done by (5.40) can be viewed as being similar to the selection of energies less than the Fermi-one in the summation over states at $T = 0$ (see also section 2.8). In this context there are reasons to say that the 1D harmonic oscillator on the lattice exhibits attributes characterizing specifically fermionic systems at $T = 0$.

5.4 Applying the factorization method

The discrete equation characterizing the Krawtchouk-polynomials

$$[x(q-1) + N_0 p]\, k_n^{(p)}(x+1, N_0) + qx k_n^{(p)}(x-1, N_0) + \tag{5.52}$$

$$+ [n - N_0 p - x(q-p)]\, k_n^{(p)}(x, N_0) = 0$$

can be rewritten, of course, as a Hamiltonian eigenvalue equation as

$$\mathcal{H} k_n^{(p)}(x, N_0) = [x(q-1) + N_0 p \exp(\partial/\partial x) + nq - x(q-p) + \tag{5.53}$$

$$+ qx \exp(-\partial/\partial x)] k_n^{(p)}(x, N_0) = p(N_0 - n) k_n^{(p)}(x, N_0) \quad .$$

A such Hamiltonian can be factorized in terms of raising and lowering operators (Lorente (2001a), Lorente (2001b)), which proceeds in a close analogy with the standard factorization method discussed long ago by Darboux (1889) and later by Infeld and Hull (1951). The interesting point is that this latter method can be generalized towards establishing a Lie-theory description of both difference (Miller (1969), Miller (1972)) and q-difference (Miller (1970)) equations.

First we have to recall that the raising operator is produced in terms of the ∇-operator , such as established before for discrete polynomials, as shown in (A.18) . The lowering operator can then be established with the help of (A.15). Indeed, one obtains

$$[qx\nabla + (N_0 - n)p - x]k_n^{(p)}(x, N_0) = -(n+1)k_{n+1}^{(p)}(x, N_0) \tag{5.54}$$

by virtue of (A.18), which shows that the raising operator can be defined as

$$L^{(+)}(x, n)k_n^{(p)}(x, N_0) \equiv [-qx\nabla - (N_0 - p) + x]k_n^{(p)}(x, N_0) \quad . \tag{5.55}$$

On the other hand, one has the recurrence relation

$$xk_n^{(p)}(x, N_0) = (n+1)k_{n+1}^{(p)}(x, N_0)+ \tag{5.56}$$

$$[p(N_0 - n) + nq]k_n^{(p)}(x, N_0) + pq(N_0 - n + 1)k_{n-1}^{(p)}(x, N_0)$$

in accord with (A.15). Inserting (5.56) into (5.55) yields

$$L^{(+)}(x, n)k_n^{(p)}(x, N_0) \equiv [qx\nabla - qn]k_n^{(p)}(x, N_0) = \tag{5.57}$$

$$= pq(N_0 - n + 1)k_{n-1}^{(p)}(x, N_0) \quad .$$

Under such circumstances one obtains

$$L^{(+)}(x, n-1)L^{(-)}(x, n)k_n^{(p)}(x, N_0) = pq(N_0 - n + 1)k_n^{(p)}(x, N_0) \tag{5.58}$$

and

$$L^{(-)}(x, n-1)L^{(+)}(x, n)k_n^{(p)}(x, N_0) = \tag{5.59}$$

$$= (pq)(n+1)(N_0 - n)k_n^{(p)}(x, N_0) \quad .$$

It is then clear that the Hamiltonian can be factorized as

$$\mathcal{H} = \frac{1}{q(n+1)}L^{(-)}(x, n+1)L^{(+)}(x, n) = \tag{5.60}$$

$$= \frac{N_0 - n}{qn(N_0 - n + 1)} L^{(-)}(x, n-1) L^{(+)}(x, n) \quad .$$

Starting from $L^{(-)}(x,0)k_0^{(p)}(x,N_0) = 0$, i.e. from $k_0^{(p)}(x,N_0) = 1$, one obtains the eigenfunction

$$k_n^{(p)}(x, N_0) = \frac{1}{n!} \prod_{l=0}^{n-1} L^{(+)}(x, l) k_0^{(p)}(x, N_0) \tag{5.61}$$

which reproduces (5.45) via $p = 1/2$ and $x \to x + N_0/2$. One proceeds similarly in the case of Krawtchouk functions (Lorente (2001b)), as already indicated by (A.22) and (A.23).

5.5 The discrete analog of the radial Coulomb-problem

Now we would like to us this opportunity in order to generalize the discrete analog of the radial Coulomb-problem , such as proposed recently in terms of normalized Meixner-functions (Lorente (2001b)). Indeed, the discrete equation of the Meixner-polynomials is given by (Nikiforov, Suslov and Uvarov (1991))

$$\mu (\gamma + x) m_n^{(\gamma,\mu)} (x+1) + x m_n^{(\gamma,\mu)} (x-1) + \tag{5.62}$$

$$+ (n - n\mu - \mu\gamma - x(1+\mu)) m_n^{(\gamma,\mu)} (x) = 0 \quad .$$

On the other hand, the normalized Meixner-function reads

$$M_n^{(\gamma,\mu)} (x) = \frac{\sqrt{\rho(x)}}{d_n} m_n^{(\gamma,\mu)} (x) = \tag{5.63}$$

$$= \sqrt{\frac{\mu^{n+x}\Gamma(\gamma + x)(1-\mu)^\gamma}{n!\Gamma(\gamma)\Gamma(x+1)(\gamma)_n}} m_n^{(\gamma,\mu)} (x)$$

which proceeds in accord with (A.14). This produces the discrete equation

$$\sqrt{\mu(x+1)(\gamma+x)} M_n^{(\gamma,\mu)} (x+1) + \sqrt{\mu x (\gamma + x - 1)} M_n^{(\gamma,\mu)} (x-1) + \tag{5.64}$$

$$+ (n - n\mu - \mu\gamma - x(1+\mu)) M_n^{(\gamma,\mu)} (x) = 0$$

for which the classical limit can be established via $x = \rho/\varepsilon$ and $\mu = 1 - \varepsilon$. So, one gets faced with the equation

$$\rho^2 \frac{d^2}{d\rho^2} \widetilde{M}_n^{(\gamma,1)}(\rho) + \rho \frac{d}{d\rho} \widetilde{M}_n^{(\gamma,1)}(\rho) + \tag{5.65}$$

$$+ \left[\rho \left(n + \frac{\gamma}{2} \right) - \frac{\rho^2}{4} - \frac{(\gamma - 1)^2}{4} \right] \widetilde{M}_n^{(\gamma,1)}(\rho) = 0$$

in which the ρ-dependence of $\widetilde{M}_n^{(\gamma,1)}(\rho)$ is selfconsistently understood. Now, the point is that (5.65) should be compared with the N-dimensional form of the classical radial Coulomb-problem, i.e. with (Nieto (1979))

$$\rho^2 \frac{d^2}{d\rho^2} R_n + (N-1)\rho \frac{d}{d\rho} R_n + \left[\lambda_1 \rho - \frac{\rho^2}{4} - l(l + N - 2) \right] R_n = 0 \tag{5.66}$$

in which $n = n_r$ stands for the radial quantum number, whereas

$$R_n(\rho) \in \left\{ L_2(0, \infty), \rho^{N-1} d\rho \right\} \quad . \tag{5.67}$$

The principal quantum number reads as $\lambda_1 = l + n_r + (N-1)/2$. Here, we also consider that the Coulomb-problem concerns the attractive $-e^2/r$ potential irrespective of N. Comparing (5.65) and (5.67), we have to perform the identifications $N = 2$ and

$$\gamma = 2l + 1 \equiv 2|m| + 1 \quad . \tag{5.68}$$

Now the principal quantum number reads

$$\lambda = n_2 = n_r + |m| + \frac{1}{2} = \frac{1}{2}, \frac{3}{2}, \ldots \tag{5.69}$$

whereas the radial wavefunction has the form

$$\widetilde{M}_{n_r}^{(\gamma,1)}(\rho) \equiv R_{n_r}^{(|m|)}(\rho) = C_1 \rho^{|m|} \exp\left(-\frac{\rho}{2} \right) L_{n_r}^{(2|m|)}(\rho) \tag{5.70}$$

where $L_{n_r}^{(2|m|}(\rho)$ denotes the related Laguerre-polynomial. The normalization constant is denoted by $C_1 = C_1(l, N)$, so that this time $C_1 = C_1(|m|, 2)$. We can proceed further towards selecting $N = 3$ by rescaling the normalized Meixner-polynomial as

$$M_n^{(\gamma,\mu)}(x) \to \mathcal{M}_{n,3}^{(\gamma,\mu)}(x) = \sqrt{x} M_n^{(\gamma,\mu)}(x) \tag{5.71}$$

where $n = n_r$ and $m = m_\varphi$. Inserting (5.71) into (5.64) yields the modified discrete equation

$$\sqrt{\mu x (\gamma + x)} \mathcal{M}_{n,3}^{(\gamma,\mu)}(x+1) + x \sqrt{\frac{\mu(\gamma + x - 1)}{x-1}} \mathcal{M}_{n,3}^{(\gamma,\mu)}(x-1) + \tag{5.72}$$

$$+ (n - \mu n - \mu\gamma - x(1+\mu)) \mathcal{M}_{n,3}^{(\gamma,\mu)}(x) = 0 \quad .$$

which produces the classical limit

$$\frac{d^2}{d\rho^2} \widetilde{\mathcal{M}}_{n,3}^{(\gamma,1)}(\rho) + \left(\rho\left(n + \frac{\gamma}{2}\right) - \frac{\rho^2}{4} - \frac{\gamma^2 - 2\gamma}{4}\right) \widetilde{\mathcal{M}}_{n,3}^{(\gamma,1)}(\rho) = 0 \quad . \tag{5.73}$$

This generates in turn the reduced radial equation for the N-dimensional Coulomb-problem :

$$\rho^2 \frac{d^2}{d\rho^2} \varphi_n(\rho) + \left(\lambda_1 \rho - \frac{\rho^2}{4} - (l + \frac{N-1}{2}\left(l + \frac{N-3}{2}\right)\right) \varphi_n(\rho) = 0 \tag{5.74}$$

where $N = 3$, such that

$$\varphi_n(\rho) = \rho^{(N-1)/2} R_n(\rho) \in \{L_2(0,\infty), d\rho\} \tag{5.75}$$

$$\gamma = 2(l+1) \tag{5.76}$$

and

$$\lambda_1 = n + l + 1 = n_r + l + 1 \tag{5.77}$$

as usual. We can then say, that the discrete analog of the three dimensional (reduced) radial Coulomb-problem is given by (5.72). A similar result has been obtained before by starting from the normalized Meixner-function (Lorente (2001b))

$$M_n^{(\gamma,\mu)}(x) \to \mathcal{M}_{L,n}^{(\gamma,\mu)}(x) = \sqrt{\mu(\gamma + x)} M_n^{(\gamma,\mu)}(x) \tag{5.78}$$

instead of (5.71). Now we are ready to say that the discrete analog of (5.66) is done by the rescaled Meixner-function

$$M_n^{(\gamma,\mu)}(x) \to \mathcal{M}_n^{(\gamma,\mu)}(x) = x^{(2-N)/2} M_n^{(\gamma,\mu)}(x) \quad . \tag{5.79}$$

The continuous limit of the discrete equation characterizing $\mathcal{M}_n^{(\gamma,\mu)}(x)$ reproduces (5.66) via

$$\gamma = 2l + N - 1 \tag{5.80}$$

and

$$\lambda_1 = n + \frac{\gamma}{2} = l + n + \frac{N-1}{2} \quad . \tag{5.81}$$

On the other hand one has the normalization

$$\mathcal{N}_{n',n} = \sum_{x=0}^{\infty} \mathcal{M}_{n'}^{(\gamma,\mu)}(x)\, \mathcal{M}_n^{(\gamma,\mu)}(x)\, x^{N-2} = \sum_{x=0}^{\infty} M_{n'}^{(\gamma,\mu)}(x)\, M_n^{(\gamma,\mu)}(x) \tag{5.82}$$

by virtue of (5.79). We shall then assume that the continuous limit

$$R_n(\rho) = \widetilde{\mathcal{M}}_n^{(\gamma,1)}(\rho) \tag{5.83}$$

is generated by virtue of the normalization

$$\mathcal{N}_{n',n} \to \int_0^{\infty} \widetilde{\mathcal{M}}_{R,n'}^{(\gamma,1)}(\rho) \widetilde{\mathcal{M}}_{R,n}^{(\gamma,1)}(\rho) \rho^{N-1} d\rho = \delta_{n',n} \tag{5.84}$$

as indicated by (5.67). One would then have

$$R_n(\rho) = R_n^{(1)}(\rho) = C_1 \rho^l \exp\left(-\frac{\rho}{2}\right) L_n^{(2L)}(\rho) \tag{5.85}$$

where $L = l + (N-2)/2$ and where the normalization constant reads

$$C_1 = C_1(l, N) = \sqrt{\frac{\Gamma(n+1)}{2\lambda_1 \Gamma(2l + N + n - 1)}} \quad . \tag{5.86}$$

5.6 The discrete analog of the isotropic harmonic oscillator

It is well known that (5.66) can be converted into the one characterizing the isotropic harmonic oscillator by performing the substitution (Kostelecky et al (1985), Kostelecky and Russell (1996))

$$\rho = r^2 \quad . \tag{5.87}$$

This proceeds in conjunction with the matching conditions

$$\frac{l_2}{l_1} = \frac{N_2 - 2}{N_1 - 1} = 2 \tag{5.88}$$

where $l_1 = l$ and $N_1 = N$, such that by now the subscript "2" concerns the isotropic oscillator implemented in this manner. Counting the solutions, there is a correspondence from on to half. However, more general maps between the solutions of initial and converted systems can also be considered, as indicated e.g. by (2.12) in Kostelecky et al (1985). Now (5.66) is converted into the radial equation

$$r^2 \frac{d^2}{dr^2} R_n^{(2)}(r) + (N_2 - 1) r \frac{d}{dr} R_n^{(2)}(r) + \tag{5.89}$$

$$+ \left[2\lambda_2 r^2 - r^4 - l_2 (l_2 + N_2 - 2)\right] R_n^{(2)}(r) = 0$$

characterizing an isotropic harmonic oscillator in N_2 space-dimensions. Now one has

$$R_n^{(2)}(r) = R_n^{(1)}(r^2) \in \left\{L_2(0, \infty), r^{N_2+1} dr\right\} \tag{5.90}$$

which shows, however, that $R_n^{(2)}(r)$ should be rescaled by a $1/r$-factor. The principal quantum number is given by

$$\lambda_2 = 2\lambda_1 = \gamma_2 + 2n = l_2 + 2n + \frac{N_2}{2} \tag{5.91}$$

whereas

$$\gamma = \gamma_1 = \gamma_2 = l_2 + \frac{N_2}{2} \quad . \tag{5.92}$$

This means in turn that (5.89) is produced, to second ε-order, by the continuous limit of the rescaled Meixner-function

$$\mathfrak{M}_n^{(Oscill)}(x) = x^{(2-N_2)/4} M_n^{(l_2+N_2/2,\mu)}(x) \tag{5.93}$$

where now

$$x = \frac{r^2}{\varepsilon} \quad . \tag{5.94}$$

Now the continuous limit of the rescaled Meixner-function

$$R_n^{(Oscill)}(r) = \widetilde{\mathfrak{M}}_n^{(\gamma,1)}\left(r^2\right) \tag{5.95}$$

is normalized from the very beginning as

$$R_n^{(Oscill)}(r) \in \left\{L_2(0,\infty),\ r^{N_2-1}dr\right\} \quad . \tag{5.96}$$

Indeed, in the present case (5.84) is replaced by

$$\mathfrak{N}_{n',n} \rightarrow \int_0^\infty \widetilde{\mathfrak{M}}_n^{(\gamma,1)}\left(r^2\right) \mathfrak{M}_n^{(\gamma,1)}\left(r^2\right) r^{N_2-1}dr = \delta_{n',n} \tag{5.97}$$

which leads in turn to (5.96). In addition, one has the nontrivial invariance property

$$\mathfrak{M}_n^{(Oscill)}(x) = \mathfrak{M}_n^{(Coul)}(x) = \mathfrak{M}_n^{(\gamma,\mu)}(x) \tag{5.98}$$

which proceeds in accord with (5.88) and (5.92). In other words, we found that duality partners just referred to above exhibit the same discrete analog. It should also be noted that

$$R_n^{(Oscill)}(r) = \frac{C_2}{C_1} R_n^{(2)}(r) \tag{5.99}$$

where

$$C_2 = C_2(l_2, N_2) = \sqrt{\frac{2\Gamma(n_r+1)}{\Gamma(l_2+n_r+N_2/2)}} \quad . \tag{5.100}$$

In particular, the discrete analog of (2.68), i.e. of the dimensionless radial equation characterizing the 2D quantum dot potential $kr_0^2/2$ is given by (5.93) under the substitutions $N_2 = 2$ and $l_2 = |m|$. This leads to the Meixner-function

$$M_n^{(QD)}(x) = M_n^{(|m|+1,\mu)}(x) \tag{5.101}$$

where $x = r^2/\varepsilon$, $m = m_\varphi$, $n = n_r$ and $\mu = 1 - \varepsilon$.

5.7 Realizations of Heisenberg-Weyl commutation relations

Our goal is to obtain informations concerning R and H just by handling the commutation relation

$$HR = qRH \tag{5.102}$$

in an algebraic manner. For this purpose let us assume that the H -and R-operators in (5.102) are tridiagonal:

$$H\psi_n = a_{n+1}\psi_{n+1} + b_n\psi_n + a_n\psi_{n-1} \tag{5.103}$$

and

$$R\psi_n = d_{n+1}\psi_{n+1} + g_n\psi_n + r_n\psi_{n-1} \tag{5.104}$$

where all the coefficients are real, such that $a_n \neq 0$ and $d_n r_n \neq 0$. The H-operator, which plays the role of the Hamiltonian, is Hermitian, as indicated by (5.103). A third operator, say $L = R^+$, can also be introduced, in which case

$$L\psi_n = r_{n+1}\psi_{n+1} + g_n\psi_n + d_n\psi_{n-1} \quad . \tag{5.105}$$

The operators H, R and L can then be viewed as generators of an $SL_q(2)$-group (Floreanini and Vinet (1995a), Floreanini and Vinet (1995b)).

Inserting (5.103) and (5.104) into (5.102) yields three kinds of matching conditions by equating the coefficients of $\psi_{n\pm 2}$, $\psi_{n\pm 1}$ and ψ_n, respectively (Zhedanov (1993)). First one has

$$a_{n+1}d_n = qa_n d_{n+1} \tag{5.106}$$

and

$$a_{n-1}r_n = qa_n r_{n-1} \quad . \tag{5.107}$$

This yields the solution

$$d_n = c_1 \frac{a_n}{q^n} \tag{5.108}$$

and

$$r_n = c_2 q^n a_n \tag{5.109}$$

where c_1 and c_2 are arbitrary real constants. In the second case we have to handle the matching conditions

$$\frac{c_1}{q^{n+1}}(b_{n+1} - qb_n) = qg_{n+1} - g_n \tag{5.110}$$

and

$$c_2 q^{n+1}(b_n - qb_{n+1}) = qg_n - g_{n+1} \quad . \tag{5.111}$$

This shows that the difference

$$(1+q)g_n - \left(\frac{c_1}{q^n} + c_2 q^{n+1}\right) b_n \equiv c_3 \tag{5.112}$$

is independent of n. We then find the coefficient b_n as

$$b_n = -c_3 \frac{qf_n + c_4}{w_n w_{n-1}} \tag{5.113}$$

in which

$$f_n = \frac{c_1}{q^n} + c_2 q^{n+1} \tag{5.114}$$

and

$$w_n = \frac{c_1}{q^n} - c_2 q^{n+2} \tag{5.115}$$

where c_4 is a further constant. In the third case one obtains the interconnection

$$v_{n+1} a_{n+1}^2 - v_{n-1} a_n^2 = \frac{q-1}{q+1}\left(f_n b_n^2 + c_3 b_n\right) \tag{5.116}$$

in which this time

$$v_n = \frac{c_1}{q^n} - c_2 q^{n+1} \quad . \tag{5.117}$$

Having established b_n such as given by (5.113), we then obtain

$$a_n = \sqrt{\frac{b_n b_{n-1} w_n w_{n-2} + c_5}{(q+1)^2 v_n v_{n-1}}} \tag{5.118}$$

by virtue of (5.116), where c_5 is an additional constant. Thus (5.102) can be solved in terms of tridiagonal representations, which results in a general solution depending on five c_i-parameters $(i = 1 - 5)$.

Now a discrete Schrödinger-equation like

$$H\psi_n = a_{n+1}\psi_{n+1} + b_n\psi_n + a_n\psi_{n-1} = E\psi_n \tag{5.119}$$

can be discussed in connection with (5.113), (5.114) and (5.118). Related potentials would then proceed in terms of typical q-dependent functions f_n, v_n and w_n quoted above. Moreover, the wavefunction

$$\varphi_n^{(m)} = R^m \psi_n \tag{5.120}$$

exhibits the eigenvalue $q^m E_n$:

$$H\varphi_n^{(m)} = q^m E_n \varphi_n^{(m)} \quad . \tag{5.121}$$

We then have to account for the orthogonality relation

$$\left\langle \psi_n \middle| \varphi_n^{(m)} \right\rangle = 0 \tag{5.122}$$

which agrees with the present selection of real $q \neq 1$ -values. On the other hand both RR^+ and R^+R are quadratic forms in H like (Zhedanov (1993))

$$RR^+ = \xi H^2 + \eta H + \zeta \tag{5.123}$$

and

$$R^+R = \xi q^2 H^2 + \eta q H + \zeta \tag{5.124}$$

where

$$\xi = \frac{c_1 c_2}{q} \tag{5.125}$$

$$\eta = -\frac{c_3 c_4}{q^2 (q+1)} \tag{5.126}$$

and

$$\zeta = \frac{c_3^2 + c_5/q}{(q+1)^2} \quad . \tag{5.127}$$

One would then obtain

$$\left\langle \varphi_n^{(1)} \middle| \varphi_n^{(1)} \right\rangle = \left(\xi q^2 E_n^2 + \eta q E_n + \zeta\right) \langle \psi_n | \psi_n \rangle \tag{5.128}$$

by virtue of (5.124), in which case the energy eigenvalues can be written down in an explicit manner as

$$E = E^{(\pm)} = \frac{\eta}{2\xi q}\left[-1 \pm \sqrt{1 + \frac{4\xi}{\eta^2}(1-\zeta')}\right] \tag{5.129}$$

provided that

$$\eta^2 \geq \frac{4\xi}{\eta^2}(\zeta' - 1) \tag{5.130}$$

where

$$\zeta' = \widetilde{\Gamma}_n \zeta \equiv \frac{\langle \psi_n | \psi_n \rangle}{\left\langle \varphi_n^{(1)} \middle| \varphi_n^{(1)} \right\rangle} \zeta \quad . \tag{5.131}$$

So, we have to realize that q-deformed algebraic structures incorporate a lot of new information, which is useful for applications in several areas (see also (9.102), or (10.126)).

Chapter 6

Hopping Hamiltonians. Electrons in Electric Field

The tight binding models are almost familiar in solid state physics (Ashcroft and Mermin (1976)). They are described specifically by so called hopping Hamiltonians , the simplest version of which looks like

$$\mathcal{H} = -t_H \sum \left(a_{i+1}^+ a_i + a_i^+ a_{i+1} \right) \tag{6.1}$$

where t_H stands for a site independent hopping parameter and where a_i^+ (a_i) creates (annihilates) a particle, e.g. an electron, at site i. For convenience, we shall assume the familiar commutation relation $[a_i, a_j^+] = \delta_{i,j}$, though the influence of fermionic anticommutator-counterparts can be discussed in a similar manner. Hoppings such as addressed in (6.1) are NN effects, but refinements concerning the insertion of next nearest neighbor terms can be easily done. In this section we shall then discuss some typical solutions and particular realizations of such interesting Hamiltonians.

The $1D$ discrete tight binding model with NN interaction such as done by the second-order discrete equation

$$t_n \psi_{n+1} + t_{n-1}^* \psi_{n-1} + V_n \psi_n = E \psi_n \tag{6.2}$$

has received much attention (Sokoloff (1985), Gredeskul et al (1997)). Here n is the site index, V_n is the n site potential and t_n is the hopping integral. Putting $t_n = t_{n-1}^* = -t_H$, yields time-independent discretized Schrödinger equations, as considered before in sections 4.1, 4.2 and 4.3. Systems for which V_n is random (Anderson (1958)), aperiodic (Sil et al (1993), Lindquist and Riklund (1998)), or quasiperiodic (Ostlund and Pandit (1984), Aubry and Andre (1980)) have been discussed before in some more detail, but periodic V_n-potentials (Harper (1955), Rauh (1975), Wannier (1975), Obermair and Wannier (1976), Hofstadter (1976)) are of special interest. If $t_n = t^*_{n-1} = t_0$ and $V_n = 0$ one deals with a free particle with the extended wavefunction

$$\psi_n = \exp(i\theta n) \tag{6.3}$$

and the energy

$$E = E(k) = 2t_0 \cos\theta \tag{6.4}$$

where $\theta = ka$. A similar θ-dependence has been invoked before in connection with the energy-band description bearing on (2.24). However, there is a common practice to use the dimensionless k-quotation instead of θ. One then says that units for which $a = 1$ are used. Under the influence of disordered potentials the wavefunction becomes localized (Anderson (1958)), but the same happens in periodic or quasiperiodic potentials for sufficiently large values of underlying gap-like parameters (Ostlund and Pandit (1984), Aubry and Andre (1980)). Besides the explicit presence of the V_n-potential, we have also to keep in mind the fact that additional interactions are able to be incorporated into the t_n and t^*_{n-1} -coefficients.

One realizes that (6.2) is produced by the two-dimensional ($2D$) Hamiltonian (Anderson (1958))

$$\mathcal{H} = \sum_i V_i |i\rangle\langle i| + \sum_{|i-j|=1} t_{ij} |i\rangle\langle j| \tag{6.5}$$

where the $|i\rangle$'s are localized orthonormalized Wannier-states for which (Nenciu (1991))

$$\langle i|\, j\rangle = \delta_{ij} \quad . \tag{6.6}$$

Then the Hamiltonian

$$\mathcal{H} = \sum_i \left(V_i |i\rangle\langle i| + t_{i,i+1} |i\rangle\langle i+1| + t_{i+1,i} |i+1\rangle\langle i|\right) \tag{6.7}$$

is Hermitian if

$$t^*_{i+1,i} = t_{i,i+1} \tag{6.8}$$

and $V_i^* = V_i$. Starting from the eigenvalue equation

$$\mathcal{H}\left|\psi\right\rangle = E'\left|\psi\right\rangle \tag{6.9}$$

and using the

$$\left|\psi\right\rangle = \sum_n C_n \left|n\right\rangle \tag{6.10}$$

reproduces (6.2) via

$$\left\langle n\right| \mathcal{H}\left|\psi\right\rangle = \mathcal{H}_n \psi_n = E'\psi_n \tag{6.11}$$

and $\psi_n = C_n$, such that

$$t_n = t_{n,n+1} = t^*_{n+1,n} \tag{6.12}$$

and

$$t_{n-1} = t_{n-1,n} = t^*_{n,n-1} \quad . \tag{6.13}$$

It should be noted, however, that Hamiltonians such as done by (6.7) can also be solved with the help of a continued fraction representation of corresponding Green-functions (Hubbard (1979)). Moreover, proofs have also been given that similar Hamiltonians are also responsible for the description of neutral atoms on $2D$ optical lattices (Jaksch and Zoller (2003), Ruuska and Törmä (2004)).

6.1 Periodic and fixed boundary conditions

Let us specify in some more detail the normalized wavefunction (6.10) as

$$\left|\psi\right\rangle = \left|\psi\left(k\right)\right\rangle = \frac{1}{\sqrt{N_s}} \sum_{n=1}^{N_s} \exp\left(iknK\right)\left|n\right\rangle \tag{6.14}$$

where now k is an integer expressing the mode-number and where K represents an additional k-independent parameter. The number of sites is denoted by N_s. Then the above wavefunction is periodic with period N_s if

$$\left\langle n\right|\left.\psi\right\rangle = \left\langle n + N_s\right|\left.\psi\right\rangle \tag{6.15}$$

in which case

$$\exp\left(ik\left(n+N_s\right)K\right)=\exp\left(iknK\right) \quad . \tag{6.16}$$

The appropriate solution is $K=2\pi/N_s$, which can be readily used in order to establish the energy. Indeed, choosing again $t_n=t_{n-1}^*=t_0$ and $V_n=V_0$, one finds the eigenvalue

$$E'=E_P\left(k\right)=V_0+2t_0\cos\left(\frac{2\pi k}{N_s}\right) \tag{6.17}$$

which relies on (4.116) and which is of interest in the description of excitons (Knox (1963)).

In the case of fixed boundary conditions , it is suitable to represent the normalized wavefunction as

$$|\varphi\rangle=|\varphi\left(k\right)\rangle=\sqrt{\frac{2}{N_s+1}}\sum_{n=1}^{N_s}\sin\left(knK\right)|n\rangle \tag{6.18}$$

which differs, of course, from (6.14) in several respects. Now, one has

$$\langle n\,|\varphi\rangle=\sqrt{\frac{2}{N_s+1}}\sin\left(knK\right) \tag{6.19}$$

so that fixed boundary conditions like

$$\langle 0\,|\varphi\rangle=\langle N_s+1\,|\varphi\rangle \tag{6.20}$$

are now fulfilled if

$$K=\frac{\pi}{N_s+1} \quad . \tag{6.21}$$

In consequence, (6.17) gets replaced by

$$E'=E_F\left(k\right)=V_0+2t_0\cos\left(\frac{\pi k}{N_s+1}\right) \quad . \tag{6.22}$$

Comparing (6.17) and (6.22), one sees immediately, that sensitivity degrees with respect to boundary conditions are increasingly lost for larger N_s-values.

6.2 Density of states and Lyapunov exponents

The density of states (DOS) plays an important role in the derivation of partition functions with the help of integrals over energies. More exactly, a sum over states is converted into the integral

$$\sum_{\{\alpha\}} f(E) = \int dE g(E) f(E) \tag{6.23}$$

where $g(E)$ stands for the DOS just referred to above. To this aim we shall begin by applying, up to subsequent rescaling factors, the formula

$$g_1(E) = \frac{1}{\pi}\left(\left|\frac{dE}{dk}\right|\right)^{-1} \tag{6.24}$$

representing the $1D$ limit of the well-known $3Dl$ DOS-formula which is familiar in solid state physics. Using e.g. (6.17), and rescaling the r.h.s. of (6.24) by $2\pi/N_s$ gives

$$g(E) = \frac{2\pi}{\pi N_s}\left|\frac{dk}{dE}\right| = \frac{1}{2\pi t_0}\left(1 - \frac{(E - V_0)}{4t_0^2}\right)^{-1/2} \tag{6.25}$$

so that the integrated DOS becomes

$$n(E) = \int_{-2t_0}^{E} g(E)\, dE = -\widetilde{n}(E)|_{-2t_0}^{E} \tag{6.26}$$

for $V_0 = 0$, where

$$\widetilde{n}(E) = -\frac{1}{\pi}\arccos\left(\frac{E}{2t_0}\right) = -\frac{2k}{N_s} \tag{6.27}$$

expresses the indefinite version of the integrated DOS.

On the other hand, the Lyapunov exponent characterizing a second-order discrete equation , can be defined as

$$\gamma(E) = \lim_{N_s\to\infty}\frac{1}{N_s}\sum_{n=0}^{N_s-1}\ln\frac{\psi_{n+1}}{\psi_n} \tag{6.28}$$

such as indicated by similar expressions proposed before by Luck (1989) and Ott (1994). We shall then obtain

$$\gamma\left(E\right)=i\frac{2\pi k}{N_s} \tag{6.29}$$

by virtue of (6.16), so that

$$\gamma_0=Re\ \left(E\right)=0 \tag{6.30}$$

and (see also Thouless (1983a))

$$Im\ \gamma\left(E\right)=-\pi\widetilde{n}\left(E\right) \quad . \tag{6.31}$$

Thus, we found that $-\pi$ times the indefinite version of the integrated DOS yields the imaginary part of the Lyapunov exponent . A such result looks intriguing, because the DOS is a well defined quantum-mechanical concept, whereas the Lyapunov exponent concerns the large time behavior of nonlinear oscillations in dynamical systems. The understanding is the iterative mapping background of (6.28). Indeed, introducing the quotient

$$R_n=\frac{\psi_{n+1}}{\psi_n} \tag{6.32}$$

one finds that (6.2) is converted, in general, into

$$t_nR_n=(E-V_n)-\frac{t^*_{n-1}}{R_{n-1}} \tag{6.33}$$

i.e. into a classical iterative mapping problem (Izrailev et al (1998); Halburd (2005)). Thus, two seemingly unrelated quantities becomes inter-related if one resorts to suitable conversions, now to the one produced by (6.32). It is understood that in more general cases, one has $\gamma_0\left(E\right)\neq 0$, which plays an important role in the definition of localized states (Aubry and Andre (1980), Sokoloff (1981), Thouless (1983a), Willkinson and Austin (1994)). Moreover, there are reasons to say that a zero Lyapunov exponent such as done by (6.30) serves as a criterion to the identification of integrable dynamical systems, too (Ramaswamy (2002)).

In general, the evaluation of Lyapunov exponents is not an easy matter (Eckmann and Ruelle (1985), Rangarajan et al (1998)). In this respect suitable approximations, the "coherent potential approximation" included (Avgin (2002)), are useful, but dispersion relations relying on analytic functions can also be considered. One would then have (see e.g. (1.6.10) and (1.6.11) in Nussenzweig (1972))

$$\gamma_0(E) = -P\int_{-\infty}^{+\infty}\frac{\widetilde{n}(E')}{E'-E}dE' \tag{6.34}$$

in accord with (6.30), so that

$$\gamma(E) = -\int_{-\infty}^{+\infty}\frac{\widetilde{n}(E')}{E'-E-i\varepsilon}dE' \tag{6.35}$$

where $\varepsilon \to 0$. Performing the integration by parts and neglecting surface terms, yields

$$\gamma(E-i\varepsilon) = \int_{-\infty}^{+\infty}\ln(E'-E)\,g(E')\,dE' \tag{6.36}$$

which is equivalent to the Thouless-formula (Thouless (1972)). Differentiating two times (5.37) one finds immediately that

$$\frac{d^2\gamma(E)}{dE^2} = \int_{-\infty}^{+\infty}\frac{g(E)}{(E'-E)^2}dE' < 0 \tag{6.37}$$

so that $\gamma(E) \equiv \gamma(E-i\varepsilon)$ is a concave function irrespective of E (Aubry and Andre (1980)).

6.3 The localization length: an illustrative example

Let us consider, as an illustrative exercise, the simplified DOS configuration

$$D(E) = \begin{cases} 1/2, & E\in[-\Delta-1,-\Delta]\bigcup[\Delta,\Delta+1] \\ 0, & otherwise \end{cases} \tag{6.38}$$

where Δ has the meaning of a gap parameter. Indeed, (6.38) describes two identical bands separated by a gap having the width 2Δ. Using (6.36) gives

$$\gamma(E) = i\frac{\pi}{2} + \frac{1}{2}\int_{\Delta}^{\Delta+1}\ln\left(E'^2 - E^2\right)dE' \tag{6.39}$$

by virtue of (6.38), so that

$$\gamma(E) = i\frac{\pi}{2} + (\Delta+1)\ln(\Delta+1) - \Delta\ln\Delta - 1+ \tag{6.40}$$

$$+E\ Arctanh\frac{E}{E^2+\Delta(\Delta+1)} \quad .$$

Assuming small E-values, such as $|E| \ll \Delta$, one finds the leading approximation

$$\gamma(E) \cong \gamma_0(E) = i\frac{\pi}{2} + \ln\Delta + \frac{1}{\Delta} \tag{6.41}$$

if $\Delta > 1$. This indicates that the present localization length exhibits a maximum at $\Delta = 1$. The existence of this length can be viewed as a signature of an insulator phase, in which case $L_{loc.} \cong 1/\ln\Delta$ (Aubry and Andre (1980), Thouless (1983a)) if $\Delta \gg 1$. Several details concerning such problems have also been discussed (Kroon et al (2002), Quieroz (2002)).

On the other hand, the asymptotic behavior of the localized state can be specified as follows

$$\psi(x) = f(x)\exp\left(-\frac{|x|}{\lambda}\right) \tag{6.42}$$

where $f(x)$ is in general a randomly varying function. Then the localization length can be identified with the exponential decay length λ displayed above. Other fingerprints of the localization can be found in the inverse participation number, i.e. in the second moment of the probability density

$$\frac{1}{P_p} = \sum|\psi(x)|^4 \tag{6.43}$$

where the wavefunction is assumed to be normalized to unity. This reproduces the volume of the system

$$P_p = V^{(N)} \equiv L^N \tag{6.44}$$

in the case of plane waves, but in general

$$\lim_{L\to\infty} P_p = L^{d^*} \tag{6.45}$$

where d^* expresses the fractal dimensionality of the system. One has $d^* \leq N$ for extended states, but d^* vanishes if the state is localized (Kramer and MacKinnon (1993)). Weak localization effects have also their own interest (Akkermans and Montambaux (2004)).

6.4 Delocalization effects

Proofs have been given that under the influence of long-range correlated disorder, the Anderson theory of localization (Anderson, 1958), which predicts an insulating behavior for $2D$ electron systems with weak disorder, fails to be valid. Indeed, in this latter case the spectral density

$$S(k) = \sum_{i=1}^{N_0/2} V_i \cos\left(\frac{2\pi i}{N_0} + \Phi_i\right) \tag{6.46}$$

of the one site energies in (6.7) is able to exhibit a power-like behavior

$$S(k) \sim \frac{1}{k^\alpha} \tag{6.47}$$

such that $\alpha > 2$ (De Moura and Lyra (1998)), where now the number of sites N_0 is even. The random phases $\Phi_i \in [0, 2\pi]$ displayed above are assumed to be independent. Then there are delocalization effects, as indicated by the zero-value of the Lyapunov exponent. This proceeds within a finite range of energies centered around $E = 0$. We then have to account for a metal-insulator transition, which invalidates scaling predictions mentioned above (Abrahams et al (1979)). Accordingly, a metallic temperature dependence has been observed in several systems like Si metal-oxide semiconductor field effect transistors (Kravchenko et al (1994)), as well as in many other $2D$ carrier systems like n-AlAs (Papadakis et al (1998)), n-GaAs (Hanein et al (1998)), or n-Si/SiGe (Lai et al (2005)).

Delocalization effects are also able to be produced by correlated impurities in $2D$ non-interacting electron systems (Hilke (2003)). For this purpose a tight-binding Anderson-model in two dimensions like

$$\psi_{l+1,j} + \psi_{l-1,j} + \psi_{l,j+1} + \psi_{l,j-1} = (E - V_{l,j})\psi_{l,j} \tag{6.48}$$

has been used. One would then obtain the extended wavefunction

$$\psi_{l,j} = \cos(\frac{l}{2}\pi)\exp(ikj) \tag{6.49}$$

and the energy $E = 2\cos k$ if $V_{2l,j} = 0$, while the $V_{2i+1,j}$'s are still assumed to be random.

Interplays between delocalization effects relying on the long-range correlated disorder and the dynamic localization implied by the presence of an electric field $\mathcal{E}$ have also been considered (Dominguez-Adame et al (2003)). For this purpose one resorts to a discrete Hamiltonian like

$$\mathcal{H} = \sum_{m=1}^{N_0} (V_m - mea\mathcal{E}_{el}) |m\rangle \langle m| + t_0 \sum_{m=1}^{N_0-1} (|m\rangle \langle m+1| + |m+1\rangle \langle m|) \quad (6.50)$$

where N_0 is again even and where V_i proceeds in accord with (6.46) and (6.47).

6.5 The influence of a time dependent electric field

The influence of a time dependent electric field, say $\mathcal{E}_{el} = E(t) = \hbar\mathcal{E}_F f(t)/ea$, on a charged particle moving on an infinite $1D$ lattice is of a special interest (Zhu et al (1999), Suqing et al (2003)). Such systems serve as a reasonable quantum-mechanical model of an $1D$ nanoconductor. The Hamiltonian (6.50) can be readily generalized towards an infinite lattice ($-\infty < i < \infty$), whereas the time-dependent version of (6.10) reads

$$| \Psi(t) >= \sum_{m=-\infty}^{\infty} C_m(t) \mid m > \quad . \quad (6.51)$$

This yields the time dependent discrete Schrödinger-equation

$$i\frac{dC_m(t)}{dt} = V(C_{m+1}(t) + C_{m-1}(t)) - m\mathcal{E}_F f(t) C_m(t) \quad (6.52)$$

by virtue of the time dependent counterpart $\mathcal{H}|\psi\rangle = i\hbar\partial|\psi\rangle/\partial t$ of (6.9), where now $V_m = 0$ and $t_0 = \hbar V$. So far one deals with an electron with the electric charge $-e < 0$, but other charged particles can be treated in a similar manner. Accounting for the continuous limit of (6.52), shows that t has the meaning of a time variable, but with the reversed sign. We can assume, for convenience, that $\mathcal{E}_F/\omega > 0$. We would like to stress that (6.52) can be exactly solved for an arbitrary time modulation $f(t)$ (Dunlap and Kenkre (1986)). In particular, a monochromatic field like $f(t) = \cos(\omega t)$ has been considered. The interesting finding in this latter reference is that there is a dynamic localization if the quotient $\mathcal{E}_F/\omega$ is a zero of the Bessel function of order zero $J_0(z)$, i.e. if

$$J_0(\mathcal{E}_F/\omega) = 0 \quad . \quad (6.53)$$

The fingerprint of this remarkable localization is the periodic return of the particle to the initially occupied site. Note that the field amplitude $\mathcal{E}_F$ looks like the frequency of Bloch oscillations predicted long ago (Bloch (1928); Citrin (2004)). Under such conditions the mean square displacement (MSD)

$$< m^2 >= \sum_{m=-\infty}^{\infty} m^2 |C_m(t)|^2 \tag{6.54}$$

remains bounded in time, which serves as a useful criterion for dynamic localization. However, we have to say that such localization effects get ruled out within the continuous limit of (6.52), as shown before (Guedes (2001)). Moreover, such effects are related to the band collapse of a quasienergy miniband (Holthaus (1992); Suqing et al (2003)). In addition, it has been found that (6.52) is of interest in the study of harmonic generation by $1D$ conductors (Pronin et al (1994), Zhao et al (1997)). The quotient $\mathcal{E}_F/\omega$ displayed above is then responsible for the maximum number of higher harmonics. Moreover, the dynamic localization such as discussed by Dunlap and Kenkre has been unambiguously confirmed in linear optical absorbtion spectra (Madureira et al (2004)). Accordingly there are firm supports to consider (6.53) as a valuable starting point towards further generalizations, as we shall see later. The influence of both electric and magnetic fields on the dynamic localization of charged particles has also been discussed (Nazareno and Brito (2001); Torres and Kunold (2004)), but such results are rather numerical.

Now we would like to show that an alternative wavefunction like

$$C_m^{(g)}(t) = \exp\left(\frac{i}{2} m(2\mathcal{E}_F\eta - \pi)\right) J_m(\Gamma) \tag{6.55}$$

can be accounted for in so far as one resorts to admissible time realizations, i.e. to a suitable time discretization (Papp (2006)). Here $J_m(z)$ denotes the Bessel function of the first kind and of order m. In addition, the complex conjugation of (6.52) gives $C_m^*(t) = \exp(i\pi m)C_{-m}(t)$, which is in accord with (6.55). Note that Bessel functions of the first kind have also been used before in the description of bound states produced by constant electric fields on the lattice, as shown by (4.36) and (4.40). Now one has

$$\Gamma = \Gamma(t) = \frac{2V}{\mathcal{E}_F f(t)} \sin(\mathcal{E}_F\eta) \tag{6.56}$$

and

$$\eta = \eta(t) = \int_0^t f(t')dt' = \int_{-t}^0 f(t')dt' \quad . \tag{6.57}$$

For this purpose one resorts to the discrete Fourier transform

$$\widetilde{C}_k(t) = \frac{1}{2\pi} \sum_{m=-\infty}^{\infty} C_m(t) \exp(-imk) \tag{6.58}$$

in which case (6.52) leads to

$$\left(\frac{\partial}{\partial t} + 2Vi\cos(k) + \mathcal{E}_F f(t) \frac{\partial}{\partial k} \right) \widetilde{C}_k(t) = 0 \quad . \tag{6.59}$$

It is understood that for macroscopically large systems the momentum can be treated as a continuous variable.

At this point let us introduce an alternative separated integration in terms of

$$\tau = \tau(t) = k + \mathcal{E}_F \eta(t) \tag{6.60}$$

instead of t. This amounts to replace $\widetilde{C}_k(t)$ by $C(s,\tau)$, where $s = t$. Accordingly, there is

$$\frac{d}{d\tau} C(s,\tau) = -\frac{2iV}{\mathcal{E}_F} \frac{\cos(\tau - \mathcal{E}_F\eta)}{f(s)} C(s,\tau) \tag{6.61}$$

which yields, for the moment, the solution

$$C(s,\tau) = c_0 \exp\left(-\frac{4iV}{\mathcal{E}_F f(s)} \sin\frac{\mathcal{E}_F\eta}{2} \cos\left(k - \frac{\mathcal{E}_F\eta}{2}\right)\right) \quad . \tag{6.62}$$

where c_0 is an integration constant. Combining (6.58) and (6.62) then leads to the alternative wavefunction, i.e. to (6.55), by invoking the rescaling $\mathcal{E}_F \to 2\mathcal{E}_F$. Indeed, applying this rescaling (see also section 6.8) yields the wavefunction

$$C_m^{(g)}(t) = \frac{1}{2\pi} \sum_{k=-\infty}^{\infty} \exp(imk - \Gamma\cos(k - \mathcal{E}_F\eta)) \tag{6.63}$$

in accord with (6.58), in which $c_0 = 1/2\pi$. This produces in turn (6.55) by virtue of the generating function of the Bessel-functions (see (8.511.4) in Gradshteyn and Ryzhik (1965))

$$\exp(iz\cos\varphi) = \sum_{m=-\infty}^{\infty} i^m J_m(z)\exp(im\varphi) \quad . \tag{6.64}$$

One sees that (6.55) obeys the initial condition

$$C_m(0) = \delta_{m,0} \tag{6.65}$$

which means, of course, that the initially occupied site is $m = 0$. It is understood again that the k-summation in (6.63) can be replaced by a related summation over the first Brillouin zone.

6.6 Discretized time and dynamic localization

Using the recurrence relation for the Bessel functions

$$2\frac{d}{dz}J_m(z) = J_{m-1}(z) - J_{m+1}(z) \tag{6.66}$$

as well as (6.56), it can be easily verified that (6.55) satisfies (6.52) if

$$\sin(\mathcal{E}_F\eta(t)) = \varepsilon' \to 0 \tag{6.67}$$

where $\mathcal{E}_F \neq 0$, which results in a nontrivial time discretization. It is clear that (6.55) leads to the actual field-free solution:

$$C_m(t) \to C_m^{(0)}(t) = \exp(-im\pi/2)J_m(2tV) \tag{6.68}$$

if $\mathcal{E}_F \to 0$. The time grid, say $t = t_g$, implemented in this manner can be established as

$$\mathcal{E}_F\eta(t_g) = (n+\varepsilon)\pi \tag{6.69}$$

where $\varepsilon = (-1)^n\varepsilon'/\pi$ and where n is an integer. Correspondingly, the MSD exhibits the limit

$$< m^2 >= \sum_{m=-\infty}^{\infty} m^2 J_m^2(\Gamma(t)) = \frac{\Gamma^2(t)}{2} \to \frac{\Gamma^2(t_g)}{2} \tag{6.70}$$

for $t \to t_g$, which proceeds in accord with the $n = 1$ form of (8.536.2) in Gradshteyn and Ryzhik (1965). The symmetry condition

$$J_{-m}(z) = (-1)^m J_m(z) \tag{6.71}$$

has also been invoked. Furthermore, there is

$$\Gamma(t_g) = \frac{2V}{\mathcal{E}_F}\frac{(-1)^n \varepsilon\pi}{f(t_g)} \tag{6.72}$$

whereas $f(t_g)$ exhibits the form

$$f(t_g) = \left[1 - \frac{\omega^2}{\mathcal{E}_F^2}(n+\varepsilon)^2\pi^2\right]^{1/2} \tag{6.73}$$

in the special case of a cosinusoidal modulation $f(t) = \cos(\omega t)$. Accordingly, the MSD becomes

$$\frac{1}{2}\Gamma^2(t_g) = \frac{2V^2}{\mathcal{E}_F^2} f_n(\varepsilon, u) \tag{6.74}$$

where $u = \mathcal{E}_F/\omega$ and

$$f_n(\varepsilon, u) = \frac{\pi^2\varepsilon^2}{1-(n+\varepsilon)^2\pi^2/u^2} \quad . \tag{6.75}$$

Now we have to realize that the characteristic function just displayed above is able to exhibit non-zero finite values like

$$f_n(\varepsilon, u) = \frac{\pi^2}{C_d} \tag{6.76}$$

if

$$1-(n+\varepsilon)^2\pi^2/u^2 = C_d\varepsilon^2 \tag{6.77}$$

and

$$u^2 = \frac{\mathcal{E}_F^2}{\omega^2} = n^2\pi^2 \tag{6.78}$$

where C_d is a positive parameter. One remarks that inserting (6.78) into (6.73) yields $f(t_g) = 0$, which can be viewed as an effective field free condition. Needless to say that a such idea is appealing from a theoretical point of view. At this point we have to say that bounded MSD-values just written down above are responsible for the onset of the dynamic localization, which proceeds in terms of (6.78). One would then have

$$\varepsilon^2(1 + C_d n^2) = -2\varepsilon n \tag{6.79}$$

so that the small ε-parameter relying on (6.69) reads

$$\varepsilon = -\frac{2n}{1 + C_d n^2} \quad . \tag{6.80}$$

Accordingly, there is $\varepsilon > 0$ ($\varepsilon < 0$) if $n < 0$ ($n > 0$), in which case $\varepsilon \to 0$ if $|n| \to \infty$. Moreover, inserting $C_d = 1$ shows that $|\varepsilon|$ decreases monotonically with $|n|$, such that $|\varepsilon| \in [0, 1]$ for $|n| \in [1, \infty)$. Equations (6.78) and (6.80) show that dynamic localization effects are able to be produced if the ratio $\mathcal{E}_F/\omega$ is centered, within small strips, around an integer multiple of π. We have to recognize, of course, that the accuracy degrees of this description get enhanced for larger $|n|$-values. This corresponds to the strong field limit, and more exactly to large values of the ratio $\mathcal{E}_F/\omega$. Accordingly, the strip-widths, i.e. the $|\varepsilon|$-magnitudes, become vanishingly small. Moreover, the leading approximation characterizing the large order zeros of $J_0(t)$ is given by $z_m \cong \pi(m - 1/4) \cong \pi m$. In this context, the present approach can also be viewed as a strong field manifestation of the one done by Dunlap and Kenkre (1986). Conversely, inserting $m - 1/4$ instead of m results in an improved approximation.

One realizes that the MSD exhibits the limit

$$< m^2 > \to 2\pi^2 \frac{V^2}{\mathcal{E}_F^2} \tag{6.81}$$

if $t \to t_g$, where $n^2 \gg 1$ and

$$C_d = 1 \quad . \tag{6.82}$$

Equation (6.81) shows that the t_g's are able to be interpreted, at least qualitatively, as the periodic return times of the particle to the initially occupied site. Combining (6.69) and (6.78) yields

$$\sin(\omega t_g) = \text{sgn}(n) \tag{6.83}$$

so that the $\varepsilon = 0$ representation of such discretized time-intervals reads

$$t_g = \frac{2\pi}{\omega}\left(n + \frac{1}{4}\mathrm{sgn}(n)\right) \tag{6.84}$$

Equation (6.84) shows that the oscillation period has the meaning of an elementary time interval.

In other words we found a reasonable strong field limit alternative to the usual wavefunction working definitely on the continuous time. The present wavefunction has been established, of course, by considering a pre-established phase, namely $m(\mathcal{E}_F\eta(t) - \pi/2)$. Indeed, this phase behaves safely under complex conjugation, whereas the field free limit proceeds in accord with (6.68). In addition, the time derivative of this phase compensates precisely the potential energy term in (6.52). This alternative is able to provide leading forms of generalized dynamical localization conditions , now by proceeding in conjunction with the implementation of a nontrivial time discretization (Papp (2006)). We can emphasize that such solutions exhibit certain similarities with the so called conditionally exactly solvable problems (see de Souza Dutra (1993)). Of course, in the present case, the former parameter fixing gets replaced by the dynamic localization condition (DLC) working in conjunction with the time discretization.

6.7 Extrapolations towards more general modulations

Eliminating t_g between

$$\sin\left(\mathcal{E}_F\eta\left(t_g\right)\right) = 0 \tag{6.85}$$

which is responsible for the $\varepsilon = 0$ -limit of (6.64) and the effective field free condition

$$f\left(t_g\right) = 0 \tag{6.86}$$

working on the time grid, provides a rather efficient proposal to the extrapolation of the strong field dynamic localization condition (DLC) (6.78) towards more general modulations of the time dependent electric field (Papp (2006)). On could emphasize, at least tentatively, that the reasoning behind this proposal is the effective field-free behavior, which can be viewed as being reminiscent to the "asymptotic" freedom behaviour characterizing high-energy descriptions in field theory. A such behavior has been checked above by the interplay between (6.84) and (6.85) when $f(t) = \cos(\omega t)$, but the incorporation of more general cases is in order. Keeping in mind that further refinements and/or explanations remain desirable, we shall then apply (6.85) and (6.86) to other signals, which results in the derivation of useful but leading approximations to DLC's one looks for.

For this purpose let us consider, as an exercise, a more general signal like

$$f(t) = f_1(t) = \cos(\omega t) + \Delta_0 \cos(2\omega t) \tag{6.87}$$

where Δ_0 is a positive parameter. Combining (6.87) with (6.85) and (6.86) then gives the DLC

$$\frac{\mathcal{E}_F}{\omega} = n\pi F_1(\Delta_0) \tag{6.88}$$

working within the strong field limit needed, where

$$F_1(\Delta_0) = \Delta_0\sqrt{8}\frac{\left[1 + \left((1+8\Delta_0^2)^{1/2} - 1\right)/4\right]^{-1}}{\left[4\Delta_0^2 - 1 + (1+8\Delta_0^2)^{1/2}\right]^{1/2}} \tag{6.89}$$

and

$$\cos(\omega t_g) = \frac{1}{4\Delta_0}\left[(1+8\Delta_0^2)^{1/2} - 1\right] \quad . \tag{6.90}$$

Now $F_1(\Delta_0)$ decreases monotonically with Δ_0, such that $F_2(\Delta_0) \in (0,1]$ for $\Delta_0 \in [0,\infty)$. In particular, (6.88) reproduces the DLC characterizing the cosinusoidal modulation as soon as $\Delta_0 \to 0$. One realizes that such results are able to serve as quickly tractable strong field limit approximations to more exact but heavily accessible DLC's. It is also clear that (6.88) works irrespective of the sign of $\mathcal{E}_F$ provided that $\omega > 0$ and $sgn(\mathcal{E}_F) = sgn(n)$. Other cases deserving further attention are presented in section 6.11. It should be stressed again that (6.85) and (6.86) stand for quickly tractable but useful approximations to DLC's for time-periodic modulations can be readily done by resorting to zero time-averages of persistent currents in L-ring circuits, as we shall see later (see (10.141)).

It should be stressed again that (6.85) and (6.86) stand for quickly tractable, but useful approximation to DLC's. However refinements concerning DLC's for time periodic modulations can be readily done by resorting to zero time-averages of persistent currents in L-ring circuits, as we shall see later (see (10.141)).

6.8 The derivation of the exact wavefunction revisited

Next we find it useful to present some details concerning the derivation of the exact wavefunction (Dunlap and Kenkre (1986))

$$C_m^{(r)}(t) = (-\lambda)^m J_m(2V|Z|) \tag{6.91}$$

characterizing (6.52). One has

$$Z \equiv U(t) + iV(t) = \exp(i\mathcal{E}_F\eta(t))Z_1(t) \quad , \tag{6.92}$$

where

$$Z_1(t) = \int_0^t dt' \exp(-i\mathcal{E}_F\eta(t')) \tag{6.93}$$

and $\lambda^2 = -Z/Z^*$. This shows that the present description depends specifically on the $\eta(t)$ -function. Inserting $d\tau = \mathcal{E}_F f(s)\, ds$ into (6.61) yields

$$\widetilde{C}_k(t) = \widetilde{C}_{k'}(0) \exp(-2iV(\cos k\mathcal{U}(t) - \sin k\mathcal{V}(t))) \tag{6.94}$$

by performing the s -integration under a fixed τ -value, i.e. under $\tau(s) = \tau(0)$. This amounts to consider the back transformation (see also (A.3) in Dunlap and Kenkre (1986))

$$k \to k' = k + \mathcal{E}_F\eta(t) \tag{6.95}$$

where by now $s = t$. Next we have to account for the Graf's addition theorem (see e.g. (8.530) in Gradshteyn and Ryzhik (1965))

$$\sum_{n=-\infty}^{\infty} J_n(2\mathcal{U}V) J_{n+\nu}(2\mathcal{V}V) = \lambda^\nu J_\nu(2V|Z|) \tag{6.96}$$

as well as for the "sine" counterpart

$$\exp(iz\sin\varphi) = \sum_{m=-\infty}^{\infty} J_m(z)\exp(im\varphi) \tag{6.97}$$

of (6.64). One would then obtain the relationship

$$C_m(t) = \sum_{\nu=-\infty}^{\infty} C_{m+\nu}(0) \exp(-i\mathcal{E}_F (m+\nu)\eta(t)) \lambda^\nu J_\nu(2V|Z|) \quad , \quad (6.98)$$

which leads to the wavefunction

$$C_m(t) = \left(-\frac{1}{\lambda}\right)^m J_m(2V|Z|) \quad (6.99)$$

by virtue of the initial condition $C_{m+\nu}(0) = \delta_{m+\nu,0}$. So far (6.99) fails, however, to satisfy (6.52). We then have to supplement the back transformation (6.95) by the rescaling

$$\mathcal{E}_F \rightarrow -\mathcal{E}_F \quad (6.100)$$

which yields in turn the exact wavefunction (6.91). We can then say that (6.100) has the meaning of a symmetry restoring condition. This means that a similar reasoning should concern the rescaling $\mathcal{E}_F \rightarrow 2\mathcal{E}_F$ used before in the derivation of (6.63).

What remains is to show in an explicit manner that the wavefunction (6.91) is actually an exact solution of (6.52). Indeed, one has

$$\frac{d}{dt}|Z| = \frac{Z+Z^*}{2|Z|} \quad (6.101)$$

so that

$$\frac{d}{dt}J_m(t) = \frac{V}{2|Z|}(Z+Z^*)(J_{m-1}(t) - J_{m+1}(t)) \quad (6.102)$$

by virtue of (6.66), where $z = 2V|Z|$. Further there is

$$mJ_m(t) = V|Z|(J_{m-1}(t) + J_{m+1}(t)) \quad (6.103)$$

in accord with (4.35), whereas

$$\lambda = i\frac{Z}{|Z|} \quad (6.104)$$

$$\frac{1}{\lambda}\frac{d\lambda}{dt} = i\mathcal{E}_F f(t) - \frac{Z-Z^*}{2|Z|^2} \quad (6.105)$$

and

$$\frac{d}{dt}(-\lambda)^m = m(-\lambda)^m \frac{1}{\lambda} \quad . \tag{6.106}$$

Accounting for (6.101)-(6.106), it can be readily verified that

$$C_m(t) = C_m^{(r)}(t) \tag{6.107}$$

fulfils precisely (6.52), as one might expect.

6.9 Time discretization approach to the minimum of the MSD

Further theoretical supports to the time discretization presented before are provided by the time minimization of the MSD

$$\langle m^2 \rangle_r = \sum_{m=-\infty}^{\infty} m^2 \left| C_m^{(r)}(t) \right|^2 = 2V^2 |Z|^2 \tag{6.108}$$

characterizing the exact wavefunction (6.91).

First one has

$$|Z|^2 = \int_0^t dt' \int_0^t dt'' \exp\left(-i\mathcal{E}_F\left(\eta(t') - \eta(t'')\right)\right) \tag{6.109}$$

by virtue of (6.92), so that

$$\frac{d}{dt}\langle m^2 \rangle_r = 4V^2 D_r(t) \tag{6.110}$$

where

$$D_r(t) = \cos\left(\mathcal{E}_F\eta(t)\right)\int_0^t dt' \cos\left(\mathcal{E}_F\eta(t')\right) + \sin\left(\mathcal{E}_F\eta(t)\right)\int_0^t dt' \sin\left(\mathcal{E}_F\eta(t')\right) \quad . \tag{6.111}$$

Resorting to the time grid $t = t_g$ implemented by (6.85), shows that

$$D_r(t_g) = 0 \tag{6.112}$$

as soon as

$$\int_0^{t_g} dt' \cos\left(\mathcal{E}_F \eta\left(t'\right)\right) = 0 \quad . \tag{6.113}$$

We then have to realize that (6.85) and (6.113) are responsible for the strong field limit of the DLC and for the exact one, respectively. Indeed, coming back to $f(t) = \cos(\omega t)$ one finds that (6.113) leads to

$$(2n+1)\frac{\pi}{2} J_0\left(\frac{\mathcal{E}_F}{\omega}\right) = 0 \tag{6.114}$$

which produces in turn the exact DLC (6.53). This proceeds by virtue of the relationship

$$(2n+1)\frac{\pi}{2} J_0\left(\frac{\mathcal{E}_F}{\omega}\right) = \int_0^{\varphi_g} d\varphi \cos\left(\frac{\mathcal{E}_F}{\omega}\sin\varphi\right) \tag{6.115}$$

where

$$\varphi_g = \omega t_g = (2n+1)\frac{\pi}{2} \tag{6.116}$$

comes from (6.84). For convenience, it has been assumed that $sgn(n) = 1$.

Differentiating once more again the MSD yields

$$\frac{d^2}{dt^2}\left\langle m^2 \right\rangle_r = 4V^2\left[1 - \mathcal{E}_F f(t) G_r(t)\right] \tag{6.117}$$

where

$$G_r(t) = \cos\left(\mathcal{E}_F \eta(t)\right)\int_0^t dt' \sin\left(\mathcal{E}_F \eta\left(t'\right)\right) - \sin\left(\mathcal{E}_F \eta(t)\right)\int_0^t dt' \cos\left(\mathcal{E}_F \eta\left(t'\right)\right) \quad . \tag{6.118}$$

Inserting $t = t_g$ gives

$$\left.\frac{d^2}{dt^2}\left\langle m^2 \right\rangle_r\right|_{t=t_g} = 4V^2 > 0 \tag{6.119}$$

which shows that the present extremal condition leads to a minimum. It is clear that the MSD-minimum established in this manner reads

$$\left\langle m^2 \right\rangle_r^{(\min)} = 2V^2 \left[\int_0^{t_g} dt' \sin\left(\mathcal{E}_F \eta\left(t'\right)\right) \right]^2 \tag{6.120}$$

in which case the pertinent DLC is produced by the interplay between (6.85) and (6.113). It is clear that the square root of the MSD-minimum displayed in (6.120) plays the role of a non-zero minimum uncertainty in the discrete position.

6.10 Other methods to the derivation of the DLC

Other methods to the derivation of the DLC have also been proposed. To this aim we have to remark that Z in (6.92) can be replaced automatically by

$$Z_1\left(t\right) = \int_0^t dt' \exp\left(-iE_F \eta\left(t'\right)\right) \tag{6.121}$$

because $|Z| = |Z_1|$. Then the MSD exhibits the form

$$\left\langle m^2 \right\rangle = 2V^2 Z_1\left(t\right) Z_1^*\left(t\right) \tag{6.122}$$

in accord with (6.108). The starting point towards establishing further descriptions is the ansatz

$$Z_1\left(t\right) = Q_1 t + Q_2\left(t\right) \tag{6.123}$$

in which Q_1 is independent of time, whereas $Q_2\left(t\right)$ oscillates with time. This means that the MSD remains bounded in time if (Suqing et al (2003))

$$Q_1 = 0 \tag{6.124}$$

which is synonymous with the DLC.

Considering as an example a time dependent electric field like

$$E\left(t\right) = E_0 + E_1 \cos\left(\omega_1 t\right) + E_2 \cos\left(\omega_2 t\right) \tag{6.125}$$

yields the field modulation

$$f(t) = \frac{\omega_B}{\mathcal{E}_F} + \cos(\omega_1 t) + \Delta \cos(\omega_2 t) \tag{6.126}$$

where $E_2 = \Delta E_1$, $\omega_B = eaE_0/\hbar$ is the usual Bloch-frequency, whereas $\mathcal{E}_F = E_1 ea/\hbar$. Accordingly, there is

$$E(t) = \frac{\hbar}{ea}\mathcal{E}_F f(t) \tag{6.127}$$

in which case

$$\mathcal{E}_F \eta(t) = \omega_B t + \alpha_1 \sin(\omega_1 t) + \alpha_2 \sin(\omega_2 t) \tag{6.128}$$

is obtained by virtue of (6.57). It is clear that $\alpha_0 = \omega_B/\mathcal{E}_F$, $\alpha_1 = \mathcal{E}_F/\omega_1$ and $\alpha_2 = \Delta\mathcal{E}_F/\omega_2$. Using (6.97), it is an easy exercise to verify that

$$Z_1(t) = \sum_{m=-\infty}^{\infty} \sum_{n=-\infty}^{\infty} J_m(\alpha_1) J_n(\alpha_2) \exp(-i\Omega_{m,n}) \frac{\sin(\Omega_{m,n} t)}{\Omega_{m,n}} \tag{6.129}$$

where

$$\Omega_{m,n} = \frac{1}{2}(\omega_B + m\omega_1 + n\omega_2) \quad . \tag{6.130}$$

Next we have to realize that the discrimination of the linear term in (6.123) can be done by restricting the sum in (6.129) to the contributions for which $\Omega_{m,n} \to 0$. This gives

$$Q_1 = Q_1(\alpha_0, \alpha_1, \alpha_2) = \sum_{m=-\infty}^{\infty} \sum_{n=-\infty}^{\infty} \left(J_m(\alpha_1)\, J_n(\alpha_2)|_{\Omega_{m,n}=0} \right) \tag{6.131}$$

which looks, however, quite involved.

Further, let us assume that $\omega_B = 0$ and $\omega_2/\omega_1 = p/q$, where p and q are mutually prime integers. Then $\Omega_{m,n} = 0$ has the roots $m = pl$ and $n = -ql$, where l is again an integer. This gives the result (Suqing et al (2003))

$$Q_1 = Q_1^{(0)} = J_0(\alpha_1) J_0(\alpha_2) + \sum_{l=1}^{\infty} J_{pl}(\alpha_1) J_{ql}(\alpha_2) \left[(-1)^{ql} + (-1)^{pl} \right] \tag{6.132}$$

which stands for a particular realization of (6.131). Moreover, there is

$$Q_1^{(0)} = Q_1^{(0)}(\alpha_1, \Delta) = J_0(\alpha_1) J_0\left(\frac{\Delta}{2}\alpha_1\right) + 2\sum_{l=1}^{\infty} J_{4l}(\alpha_1) J_{2l}\left(\frac{\Delta}{2}\alpha_1\right) \tag{6.133}$$

if $p = 2$ and $q = 1$. This means that the "exact" but involved counterpart of the DLC (6.88) is given by

$$Q_1^{(0)}(\alpha_1, \Delta) = 0 \tag{6.134}$$

but a such equation remains again too hardly tractable to be useful for explicit calculations in practice. We then have to resort to numerical methods, but explicit summations in (6.132) and (6.133) remain, if at all, to be done. In addition, there are reasons to say that the ansatz (6.123) is not generally valid, as there are systems characterized e.g. by a cubic growth of the MSD (Hufnagel et al (2001)).

6.11 Rectangular wave fields and other generalizations

Our next task is to discuss the localization attributes characterizing a periodic rectangular wave field like

$$\mathcal{E}(t) = \begin{cases} \mathcal{E}_F \,, & t \in (0, T/2) \\ -\mathcal{E}_F \,, & t \in (T/2, T) \end{cases} \tag{6.135}$$

where $E(t) = \mathcal{E}(t) W/ea$, such that $\mathcal{E}(t) = \mathcal{E}(t+T)$. A modified version of (6.135) has also been discussed before (Zhu et al (1999)). The above field exhibits sign-change discontinuities of the first kind at

$$t = t_{dis} = n_{dis}\frac{T}{2} \tag{6.136}$$

where n_{dis} is an arbitrary integer, while T denotes the period measured in units of $\hbar/W$. Equation (6.136) relies on the "exact" area quantization condition proposed before (Dignam and de Sterke (2002)). This means that the total area under the $\mathcal{E}(t)$-curve between two discontinuities encompassing an integer number of periods should be an integral multiple of 2π. Now, the total area for one period is zero, which indicates that we have to consider reasonably a half period. One would then obtain the area

$$\mathcal{E}_F\frac{T}{2} = \mathcal{E}_F\frac{\pi}{\omega} = 2n_a\pi \tag{6.137}$$

when the discontinuities are located at $n_{dis} = 0$ and $n_{dis} = 1$, where n_a is a non-zero integer.

One realizes that the periodic function characterizing (6.135) is given by the Fourier-series

$$f(t) = \frac{4}{\pi} \sum_{n=0}^{\infty} \frac{\sin(2n+1)\omega t}{(2n+1)} \tag{6.138}$$

so that $\mathcal{E}(t) = \mathcal{E}_F f(t)$, as usual. This gives

$$\eta(t) = \frac{4}{\pi\omega} \left[\frac{\pi^2}{8} - \sum_{n=0}^{\infty} \frac{\cos(2n+1)\omega t}{(2n+1)^2} \right] \tag{6.139}$$

in accord with (6.57). Applying (6.85) and (6.86) one finds

$$\omega t_g = \frac{2\pi}{T} t_g = n'\pi \quad . \tag{6.140}$$

which leads to a DLC like

$$\frac{\mathcal{E}_F}{\omega} = n \quad . \tag{6.141}$$

if n' is an odd integer. The series

$$\sum_{n=0}^{\infty} \frac{1}{(2n+1)^2} = \frac{\pi^2}{8} \tag{6.142}$$

has also been used. Now it is clear that (6.141) reproduces precisely (6.137) as soon as $n = 2n_a$, which means that we have to account for even realizations of n in (6.69). Accordingly, the periodicity of the modulation gets accounted for, too. This provides further theoretical supports favoring the accuracy attributes of the present approach to the derivation of DLC's.

The above results can be readily generalized towards the modulation

$$f(t) = f_1(t) = \sum_{n=0}^{\infty} b_{2n+1} \sin((2n+1)\omega t) \tag{6.143}$$

where $\mathcal{E}(t) = \mathcal{E}_F f(t)$, $\omega = 2\pi/T$ and $f_1(t) = f_1(t+T)$. Now (6.140) is preserved, in which case the DLC reads

$$\frac{2\mathcal{E}_F}{\omega}\sum_{n=0}^{\infty}\frac{b_{2n+1}}{2n+1}=n\pi \tag{6.144}$$

instead of (6.141), where the n' -values needed are again odd integers.

Proceeding in a similar manner one finds that the DLC's characterizing the modulations

$$f(t)=f_2(t)=\sum_{n=1}^{\infty}b_{2n}\sin\left(2n\omega t\right) \tag{6.145}$$

and

$$f(t)=f_3(t)=\sum_{n=0}^{\infty}b_{2n+1}\cos\left(\left(2n+1\right)\omega t\right) \tag{6.146}$$

are given by

$$\frac{\mathcal{E}_F}{\omega}\sum_{m=0}^{\infty}\frac{b_{4m+2}}{2m+1}=n\pi \tag{6.147}$$

and

$$\frac{\mathcal{E}_F}{\omega}\sum_{m=0}^{\infty}(-1)^m\frac{b_{2m+1}}{2m+1}=(-1)^{n'}n\pi=\pm n\pi \tag{6.148}$$

respectively, which may be useful in several respects. The time grid reads $\omega t_g=(2n'+1)\pi/2$ in both cases. We can assume, without loss of generality, that n' in (6.148) is an even integer.

Besides the area quantization condition mentioned above, the "exact" dynamic localization proceeds in terms of a further condition like (Dignam and Sterke (2002), Domachuk et al (2002), Wan et al (2004))

$$\beta_p\left(T\right)=\int_0^T\exp\left(-ip\mathcal{E}_F\eta\left(t\right)\right)dt=0 \tag{6.149}$$

where p is a non-zero integer and where T denotes the period characterizing the $f\left(t\right)$ modulation. The "exact" dynamic localization would then occur only when $\beta_p\left(T\right)=0$ for all p's, which is unlikely to be fulfilled in practice. Inserting e.g. $f\left(t\right)=\cos\left(\omega t\right)$ into (6.149) gives $\beta_p\left(T\right)=TJ_0\left(p\mathcal{E}_F/\omega\right)=0$, which shows that the "exact" description in the sense just referred to above is not possible. Related approximate versions to the dynamic localization have also been proposed. However, such approximations are too involved and concern again a rather restricted class of modulations only (Domachuk et al (2002)).

6.12 Wannier-Stark ladders

In a constant electric field the stationary version of (6.52) is given by

$$V\left(C_{m+1}+C_{m-1}\right)-m\mathcal{E}_F C_m=EC_m \tag{6.150}$$

in which case the dimensionfull energy is given, up to the sign, by $E_0=\hbar E$. This equation has also been discussed from a standard quantum-mechanical point of view in section 4.2, as indicated by (4.28)-(4.40). The present quotations rely on the ones used in section 4.2 via $\mathcal{E}_F=\mathcal{F}$, $V=t_H$ and $C_m=\psi_m$. Now one looks for a suitable solution like

$$C_m=\int_{-\pi}^{\pi}dk\exp\left(imk\right)\widetilde{C}\left(k\right) \tag{6.151}$$

where this time the integral runs over the first Brillouin zone instead of $k\in(-\infty,\infty)$. The dimensionless wave number is denoted by k, so that $k\in[-\pi,\pi]$. Inserting (6.153) into (6.152) yields the first-order differential equation

$$i\frac{d}{dk}\ln\widetilde{C}\left(k\right)=\frac{2V}{\mathcal{E}_F}\cos k-\frac{E}{\mathcal{E}_F} \tag{6.152}$$

which is the stationary counterpart of (6.59). Equation (6.154) can be immediately integrated as

$$\widetilde{C}\left(k\right)=\widetilde{C}\left(0\right)\exp\left[i\left(\frac{E}{\mathcal{E}_F}k-\frac{2V}{\mathcal{E}_F}\sin k\right)\right] \quad . \tag{6.153}$$

Next we have to apply (6.97), in which case the solution one looks for is given by

$$C_m=2\pi\widetilde{C}\left(0\right)J_{m+E/\mathcal{E}_F}\left(\frac{2V}{\mathcal{E}_F}\right) \tag{6.154}$$

which works in conjunction with the orthogonality condition

$$\sum_{k=-\pi}^{\pi} \exp ik \left(m + m' + E/\mathcal{E}_F\right) = \delta_{m',-m-E/\mathcal{E}_F} \quad . \tag{6.155}$$

For this purpose the quotient $E/\mathcal{E}_F$ must be an integer, which is responsible in turn for the implementation of equally spaced energy levels like

$$E = E_n = n\mathcal{E}_F = n\omega_B \tag{6.156}$$

where $n = 0, \pm 1, \pm 2, \ldots$. One sees that (6.158) produces an actual generalization of (4.46). On the other hand, one has the orthogonality property (see e.g. Wacker (1998))

$$\sum_{m=-\infty}^{\infty} J_{m+h}(z) J_m(z) = \delta_{k,0} \tag{6.157}$$

where z is real. Inserting $z = 2V/\mathcal{E}_F$ and using (6.158) produces the normalization condition

$$\sum_{m=-\infty}^{\infty} \left|C_m\right|^2 = 1 \tag{6.158}$$

as soon as $2\pi \mid \widetilde{C}(0) \mid = 1$. So, the discrete energies established in accord with (6.158) concern actually localized bound states. Such energies have been referred to as Wannier-Stark ladders (Wannier (1960), Fukuyama (1973), Bleuse (1988)). Wannier-Stark resonances have also been discussed (Nenciu (1991)), which also means, of course, that the level width should be smaller than the level spacing. We would like to mention, quite satisfactorily, that such ladders have been found in photoluminescence experiments with GaAs/GaAlAs superlattices (Voisin et al (1988)).

6.13 Quasi-energy approach to DLC's

There is still a useful method which can be used in the derivation of DLC's, namely the quasi-energy description. To this aim we have to assume again that the external electric field is periodic in time with period T. Then the time dependent Schrödinger-equation

$$\mathcal{H}(x,t)\psi(x,t) = i\frac{\partial}{\partial t}\psi(x,t) \tag{6.159}$$

where $x = ma$ and $\hbar = 1$ can be handled in terms of the Floquet-factorization (see e.g. Holthaus and Hone (1993))

$$\psi(x,t) = \exp(-iEt)u_E(x,t) \tag{6.160}$$

where $\mathcal{H}(x,t) = \mathcal{H}(x,t+T)$, E denotes the corresponding quasi-energy and

$$u_E(x,t) \equiv < x|t, E >= u_E(x,t+T) \quad . \tag{6.161}$$

This gives the eigenvalue equation

$$\left(\mathcal{H}(x,t) - i\frac{\partial}{\partial t}\right) u_E(x,t) = E u_E(x,t) \tag{6.162}$$

which has to be solved by accounting for the periodicity condition (6.161). Applying once more again the orthonormalized Wannier basis (6.6):

$$|t, E >= \sum_{m=-\infty}^{\infty} c_m(t)|m > \tag{6.163}$$

yields the equation

$$i\frac{d}{dt}c_m(t) = \sum_{m'=-\infty}^{\infty} < 0|\mathcal{H}|m' > c_{m+m'}(t) - (m\mathcal{E}_F f(t) + E)c_m(t) \tag{6.164}$$

which relies on (6.52). It is understood that $\mathcal{H}_0$ stands for the time independent part of the Hamiltonian, in which case there is

$$< m|\mathcal{H}_0|m' >=< 0|\mathcal{H}_0|m - m' > \tag{6.165}$$

by virtue of the translation symmetry. Accordingly, the periodic energy dispersion law of the parent band is given by

$$E_0(k) = E_0(k+2\pi) = \sum_{m=-\infty}^{\infty} < 0|\mathcal{H}_0|m > \exp(imk) \quad , \tag{6.166}$$

which leads to the well known expression (see also (6.4))

$$E_0(k) = 2V \cos k \tag{6.167}$$

by virtue of (6.52).

Next let us perform the gauge transformation

$$c_m(t) = \exp[i(Et + m\mathcal{E}_F\eta(t))]C_m(t) \quad , \tag{6.168}$$

in accord with (6.57). Then (6.164) becomes

$$i\frac{d}{dt}C_m(t) = \sum_{m'=-\infty}^{\infty} < 0|\mathcal{H}_0|m' > \exp(im'\mathcal{E}_F\eta(t))C_{m+m'}(t) \quad . \tag{6.169}$$

At this point we have to apply the discrete Fourier-transform (6.151). Equation (6.169) can then be converted as

$$i\frac{d}{dt}\widetilde{C}_k(t) = E_0(k + \mathcal{E}_F\eta(t))\widetilde{C}_k(t) \quad , \tag{6.170}$$

where

$$\widetilde{C}_{k+2\pi}(t) = \widetilde{C}_k(t) \quad . \tag{6.171}$$

Equation (6.170) can be readily integrated as

$$\widetilde{C}_k(t) = \widetilde{C}_k(0) \exp\left(-i\int_0^t dt' E_0(k + \mathcal{E}_F\eta(t'))\right) \quad , \tag{6.172}$$

whereas the periodicity condition (6.161) leads to

$$C_m(T) = C_m(0) \exp[-i(ET + m\mathcal{E}_F\eta(T))] \quad . \tag{6.173}$$

So one obtains

$$\widetilde{C}_k(T) = \widetilde{C}_{k+\mathcal{E}_F\eta(T)}(0) \exp(-iET) \quad , \tag{6.174}$$

which produces in turn the quasi-energy quantization condition

$$\widetilde{C}_{k+\mathcal{E}_F\eta(T)}(0) \exp(-iET) = \widetilde{C}_k(0) \exp\left(-i\int_0^T dt' E_0(k + \mathcal{E}_F\eta(t'))\right) \quad . \tag{6.175}$$

In the pure ac-case one has $\eta(T) = 0$, so that the corresponding quasi-energy reads

$$E = E(k, n') = \frac{1}{T} \int_0^T dt' E_0(k + \mathcal{E}_F \eta(t')) + \frac{2\pi n'}{T} \tag{6.176}$$

where n' is a further integer. Coming back to the cosinusoidal modulation $f(t) = \cos(\omega t)$ then gives the quasi-energy

$$E(k, 0) = 2V J_0 \left(\frac{\mathcal{E}_F}{\omega} \right) \cos k \tag{6.177}$$

which proceeds in accord with (96.97) and (6.167). This shows that there is $E(k, 0) = 0$ irrespective of k in so far as (6.53) is fulfilled, which is responsible for the so called band-collapse condition discussed before (Holthaus (1992)).

6.14 The quasi-energy description of dc-ac fields

The influence of the dc-ac electric field superposition

$$E(t) = E_0 + E_1 \cos(\omega t) \tag{6.178}$$

is a special case which has received much interest (Zhao et al (1995), Glück et al (1998), Zhang et al (2002), Suqing et al (2003)). Now one has

$$f(t) = \frac{\omega_B}{\mathcal{E}_F} + \cos(\omega t) \tag{6.179}$$

and

$$\mathcal{E}_F \eta(t) = \omega_B t + \alpha_1 \sin(\omega t) \tag{6.180}$$

where $\alpha_1 = \mathcal{E}_F/\omega$, which is in accord with the $E_2 = 0$ and $\omega_1 = \omega$ form of (6.125)-(6.128). Accounting for (6.171), one sees immediately that the quasi-energy formula (6.176) is preserved as soon as the angle $\omega_B T$ is an integer multiple of 2π: $\omega_B T = \mathcal{E}_F \eta(T) = 2\pi \omega_B/\omega = 2n\pi$. Then the dynamic localization occurs by virtue of the condition (Zhao et al (1995)).

$$J_n \left(\frac{\mathcal{E}_F}{\omega} \right) = 0 \quad , \tag{6.181}$$

which has the meaning of an $n \neq 0$ counterpart of (6.53).

The next point is to discuss the dynamic localization when the n integer in (6.181) is replaced by a rational number such as given by

$$\omega_B = \frac{P}{Q}\omega \tag{6.182}$$

where P and Q are mutually prime integers. This time there is

$$\widetilde{C}_{k+2\pi P/Q}(0)\exp(-iET) = \widetilde{C}_k(0)\exp\left(-i\int_0^T dt' E_0(k+\mathcal{E}_F\widetilde{\eta}(t'))\right) \tag{6.183}$$

by virtue of (6.175), where

$$\mathcal{E}_F\widetilde{\eta}(t) = \frac{P}{Q}\omega t + \alpha_1 \sin(\omega t) \quad . \tag{6.184}$$

In order to proceed further we have to apply an extra wavenumber discretization such is given by

$$k = s + \frac{2\pi}{Q}l \quad . \tag{6.185}$$

One has $k \in [-\pi, \pi]$, $s \in [-\pi/Q, \pi/Q]$ and $l = 0, 1, ..., Q-1$. This also means that

$$\Delta k = \Delta s + \frac{2\pi}{Q}\Delta l = 2\pi \tag{6.186}$$

so that parameter intervals just displayed above work safely. Inserting successively the selected l-values into (6.183) and multiplying the Q equations obtained in this manner yields the quasi-energy as

$$E_Q(s,n') = \frac{1}{QT}\sum_{l=0}^{Q-1}\int_0^T dt' E_0(s + \frac{2\pi}{Q}l + \mathcal{E}_F\widetilde{\eta}(t')) + \frac{2\pi n'}{QT} \tag{6.187}$$

instead of (6.176), which proceeds in accord with (6.171). Using (6.166) it can be easily verified that only terms for which $m = Qj$ survive the l-summation, where j is an integer. Then (6.187) can be converted as

$$E_Q(s,n') = \sum_j < 0|H_0|Qj > J_{Pj}(Qj\mathcal{E}_F/\omega)\exp(ij(Qs+\pi P)) + \frac{\omega_B n'}{P} \tag{6.188}$$

which exhibits real values as soon as H_0 is Hermitian. Checking numerically (6.188), it has been found that the dynamic localization can occur irrespective of $\mathcal{E}_F/\omega$ (Zhao et al (1995), Zhang et al (2002)), but the derivation of further analytical details is still an open problem. Quasi-enery corrections going beyond the NN-description have also been discussed (Jivulescu and Papp (2006)).

6.15 Establishing currents in terms of the Boltzmann equation

The motion of electrons in superlattices under a time dependent electric field $E(t)$ can also be treated by resorting to the Boltzmann equation in the relaxation time approximation (Ashcroft and Mermin (1976))

$$\frac{\partial}{\partial t} f(k,t) - \frac{e}{\hbar} E(t) \frac{\partial}{\partial k} f(k,t) = -\frac{1}{\tau(k)} \left[f(k,t) - f_D(k) \right] \tag{6.189}$$

The relaxation time, which is responsible for the incorporation of scattering effects is denoted by $\tau(k)$, k is the wavenumber, while $f_D(k)$ stands for the Fermi-Dirac distribution. For convenience we shall assume that the relaxation time is a constant, which amounts to say that $\tau(k) = \tau_0$. Applying again a dc-ac electric field like (6.178), yields the distribution function

$$f(k,t) = \int_{-\infty}^{t} \frac{dt'}{\tau_0} \exp\left(-\frac{t-t'}{\tau_0}\right) f_D(k+k_1) \tag{6.190}$$

where

$$k_1 = k_1(t,t') = \frac{\mathcal{E}_F}{a} \left(\eta(t) - \eta(t')\right) \tag{6.191}$$

proceeds in accord with (6.57) and (6.127). We shall also consider a general energy dispersion law

$$E_d(k) = \sum_{n=0}^{\infty} R_n \cos(kna) \tag{6.192}$$

which accounts for effects going beyond the NN-description.

Equation (6.192) provides the velocity

$$v(k) = \frac{1}{\hbar}\frac{\partial}{\partial k}E_d(k) = -\frac{a}{\hbar}\sum_{n=1}^{\infty} nR_n \sin(kna) \tag{6.193}$$

which generates in turn the time dependent current as follows

$$J(t) = -e < v(k) >_f \tag{6.194}$$

where the average has to be performed in terms of the non-equilibrium distribution function $f(k,t)$. At this point we can invoke the limits of low temperature ($T \to 0$) and low densities ($E_F \to 0$), which results in appreciable simplifications. Indeed, under such conditions one has

$$f_D(k+k_1) = \left[\exp\left(\frac{\hbar^2(k+k_1)^2}{2m_e} - E_F\right)k_BT + 1\right]^{-1} \cong 1 \tag{6.195}$$

if $k + k_1 \cong 0$, so that electrons get concentrated around $k \cong -k_1$. Accordingly, $J(t)$ becomes

$$J(t) \cong J_0 \sum_{n=1}^{\infty} \frac{nR_n}{R_1} \int_0^{\infty} \frac{ds}{\tau_0} \exp\left(-\frac{s}{\tau_0}\right) \times \tag{6.196}$$

$$\sin\left[n\omega_B s + n\frac{\mathcal{E}_F}{\omega}\left(\sin(\omega t) - \sin(\omega(t-s))\right)\right]$$

where $J_0 = eaR_1/\hbar$. Using (6.64) and (6.97), one finds the expression

$$J(t) = J_0 \sum_{n=1}^{\infty} \sum_{\nu=-\infty}^{\infty} \sum_{m=-\infty}^{\infty} \frac{nR_n}{R_1} J_\nu\left(n\frac{\mathcal{E}_F}{\omega}\right) g_{\nu n}(t) \tag{6.197}$$

where

$$g_{\nu n}(t) = Re\left[\frac{\exp i(\nu\omega t + (n\mathcal{E}_F/\omega)\sin(\omega t))}{n\omega_B + \nu\omega + i/\tau_0}\right] \tag{6.198}$$

Such results open the way for further investigations, with a special emphasis on photon assisted transport properties, band collapse effects and the generation of higher harmonics (Zhao et al (1997)).

Chapter 7

Tight Binding Descriptions in the Presence of the Magnetic Field

Under the influence of the magnetic field, the general hopping Hamiltonian of a $2D$ lattice gets specified as (Fradkin (1991), Kohmoto (1989), Peter et al (1989), Kohmoto and Hatsugai (1990), Hatsugai and Kohmoto (1990), Hatsugai (1993), Honda and Tokihiro (1997))

$$\mathcal{H} = \sum_{i,j} t_{ij} a_j^+ a_i \exp\left(i\theta_{ij}\right) + H.c. \tag{7.1}$$

where t_{ij} denotes, as before, the hopping integral between sites i and j. The electron annihilation (creation) operator at site i is denoted by a_i (a_i^+), whereas the term "$H.c$" stands for the Hermitian conjugation. The magnetic phase factor is defined by

$$\theta_{ij} = \frac{2\pi}{\Phi_0} \int_i^j \vec{A} \cdot d\vec{l} \tag{7.2}$$

where $\vec{A}$ is the vector potential. The magnetic flux quantum has been denoted by $\Phi_0 = hc/e$, as usual. Choosing, for convenience, a homogenous and transversal magnetic field, one works in many cases in the Landau-gauge $\vec{A} = (0, Bx, 0)$, but other gauges are also of interest. The sites i and j being located in a plane produce specifically square, hexagonal, triangular, or other kinds of lattices. Links between i and j sites are also specified in terms of corresponding $d\vec{l}$ -displacements, where $dl = \sqrt{dx^2 + dy^2}$. The magnetic flux through a plaquette is then given by

$$\Phi = \int \vec{B} \cdot d\vec{S} = \sum_{plq.} \int_i^j \vec{A} \cdot d\vec{l} = \frac{\Phi_0}{2\pi} \sum_{plq.} \theta_{ji} \quad . \tag{7.3}$$

The quotient $\beta = \Phi/\Phi_0$, which expresses the number of flux quanta per unit cell, has the meaning of a commensurability parameter. We shall focus our attention on rational values like

$$\beta = \frac{\Phi}{\Phi_0} = \frac{P}{Q} \in [0,1] \tag{7.4}$$

of this quotient, where P and Q are mutually prime integers, which proceeds in a close analogy with (2.51). One has e.g.

$$\beta \equiv \frac{\hbar^*}{2\pi} = \frac{eBa^2}{hc} \in [0,1] \tag{7.5}$$

for the square lattice, in which $\hbar^*$ plays the role of a cyclic commensurability parameter (see also Wilkinson (1986)).

7.1 The influence of the nearest and next nearest neighbors

Along the x-direction one has the NN links between the initial points

$$i = (n,m) \quad and \quad (n, m+1) \tag{7.6}$$

and the final points

$$j = (n+1,m) \quad and \quad (n+1, m+1) \tag{7.7}$$

respectively, for which $t_{ij} = t_a$. The same happens along the y-direction for

$$i = (n,m) \quad and \quad (n+1, m) \quad , \tag{7.8}$$

and

$$j = (n,m+1) \quad and \quad (n+1, m+1) \quad , \tag{7.9}$$

respectively, but now $t_{ij} = t_b$. The next nearest neighbors (NNN) act along bisectrices, such that $t = t_c$ ($t = t_d$) for $i = (n,m)$ and $j = (n+1, m+1)$ $i = (n+1,m)$ and $j = (n, m+1)$). Inserting $t_d = 0$ one deals with a triangular lattice. For convenience a square lattice has been assumed ($a = b$), but one proceeds similarly if $a \neq b$. The corresponding θ_{ij}-phases can then be easily evaluated, as indicated in Table 7.1.

Table 7.1. The θ_{ij}-phases in the Landau gauge for NN -and NNN -links on the square lattice. The corresponding $d\vec{l}$-displacements are also inserted.

i	j	θ_{ij}	$d\vec{l}$
(n,m)	$(n+1,m)$	0	$\vec{e}_1 dx$
$(n,m+1)$	$(n+1,m+1)$	0	$\vec{e}_1 dx$
(n,m)	$(n+1,m+1)$	$2\pi\beta\,(n+1/2)$	$(\vec{e}_1+\vec{e}_2)\,dl/\sqrt{2}$
(n,m)	$(n,m+1)$	$2\pi\beta n$	$\vec{e}_2 dy$
$(n+1,m)$	$(n+1,m+1)$	$2\pi\beta\,(n+1)$	$\vec{e}_2 dy$
$(n+1,m)$	$(n,m+1)$	$2\pi\beta\,(n+1/2)$	$(\vec{e}_2-\vec{e}_1)\,dl/\sqrt{2}$

The eigenvalue equation

$$\mathcal{H}\,|\varphi\rangle = E\,|\varphi\rangle \tag{7.10}$$

can be rewritten in the double discrete (n,m)-representation as

$$\langle n,m|\,\mathcal{H}\,|\varphi\rangle = \mathcal{H}_{n,m}\varphi_{n,m} = E\varphi_{n,m} \tag{7.11}$$

where

$$|n,m\rangle = a_n^+ a_m^+\,|0\rangle \quad . \tag{7.12}$$

Accounting for hopping integrals and using the pure-phase parameter

$$q = \exp\left(i\pi\beta\right) = \exp\left(i\pi\frac{P}{Q}\right) \tag{7.13}$$

one finds the explicit realization of (7.11) as

$$\mathcal{H}_{n,m}\varphi_{n,m} = t_a\left(\varphi_{n-1,m}+\varphi_{n+1,m}\right) + t_b\left(\varphi_{n,m-1}q^{2n}+\varphi_{n,m+1}q^{-2n}\right) + \tag{7.14}$$

$$+\frac{t_c}{q}\left(\varphi_{n-1,m-1}q^{2n}+\varphi_{n+1,m+1}q^{-2n}\right) + t_d q\left(\varphi_{n+1,m-1}q^{2n}+\varphi_{n-1,m+1}q^{-2n}\right)$$

$$= E\varphi_{n,m} \quad .$$

Of course, the hopping integrals work irrespective of the unit cell, whereas q such as given by (7.13) is a root of unity:

$$q^{2Q} = 1 \quad . \tag{7.15}$$

This means that $\varphi_{n,m}$ is a double periodic function with period Q:

$$\varphi_{n+Q,m+Q} = \varphi_{n,m} \tag{7.16}$$

which is useful for further calculations.

7.2 Transition to the wavevector representation

Let us consider the Fourier-transform

$$\varphi_{n,m} = \frac{1}{(2\pi)^2} \int_0^{2\pi} d\theta_1 \int_0^{2\pi} d\theta_2 \exp\left(i\left(n\theta_1 + m\theta_2\right)\right) \widetilde{\varphi}\left(\overrightarrow{\theta}\right) \tag{7.17}$$

in which the wave-vector is restricted to a single Brillouin zone , i.e. $\theta_i = k_i a \in [0, 2\pi]$. Then (7.14) becomes

$$\begin{aligned} \widetilde{\mathcal{H}}\widetilde{\varphi}\left(\overrightarrow{\theta}\right) &= 2t_a \cos\theta_1 \widetilde{\varphi}(\overrightarrow{\theta}) + t_b \left(\omega_2 \cos\left(\theta_1 + 2\pi\beta, \theta_2\right) + \frac{1}{\omega_2}\widetilde{\varphi}\left(\theta_1 - 2\pi\beta, \theta_2\right)\right) \\ &+ t_c q \left(\omega_1\omega_2 \widetilde{\varphi}\left(\theta_1 + 2\pi\beta, \theta_2\right) + \frac{1}{\omega_1\omega_2}\widetilde{\varphi}\left(\theta_1 - 2\pi\beta, \theta_2\right)\right) + \\ &+ \frac{t_d}{q}\left(\frac{\omega_2}{\omega_1}\widetilde{\varphi}\left(\theta_1 + 2\pi\beta, \theta_2\right) + \frac{\omega_1}{\omega_2}\widetilde{\varphi}\left(\theta_1 - 2\pi\beta, \theta_2\right)\right) \end{aligned} \tag{7.18}$$

where $\omega_l = \exp\left(i\theta_l\right)$ and $l = 1, 2$. One sees that the above terms are not sensitive with respect to θ_2, so that it can be hereafter ignored. Now, we are ready to consider an extra discretization of the θ_1-coordinate as (Hatsugai and Kohmoto (1990))

$$\theta_1 = \theta_{10} + 2\pi\beta j \tag{7.19}$$

such that (see also (6.185))

$$\int_0^{2\pi} d\theta_1 = 2\pi = \sum_{j=1}^{Q} \int_0^{2\lambda} d\theta_{10} \quad , \tag{7.20}$$

where $j = 1, 2, \ldots, Q$ and where λ remains to be determined in a selfconsistent manner. This yields

$$\lambda = \frac{\pi}{Q} \tag{7.21}$$

so that the "residual" Brillouin zone becomes Q-times narrower. On the other hand one has

$$\theta_1 \pm 2\pi\beta = \theta_{10} + 2\pi\beta\,(j \pm 1) \tag{7.22}$$

by virtue of (7.19), which leads to

$$\widetilde{\varphi}\,(\theta_1 \pm 2\pi\beta) = u_{j\pm1}\,(\theta_{10}) \tag{7.23}$$

as soon as one makes the identification

$$\widetilde{\varphi}\,(\theta_1) \equiv u_j\,(\theta_{10}) \quad . \tag{7.24}$$

Under such circumstances the eigenvalue equation becomes

$$\widetilde{B}^*_{j-1} u_{j-1} + A_j u_j + \widetilde{B}_j u_{j+1} = E u_j \tag{7.25}$$

where

$$A_j = 2t_a \cos\left(\theta_{10} + 2\pi\beta j\right) \tag{7.26}$$

and

$$\widetilde{B}_j = \omega_2 \left(t_b + \omega_{10} t_c q^{2j+1} + \frac{1}{\omega_{10}} \frac{t_d}{q^{2j+1}} \right) \tag{7.27}$$

where $\omega_{10} = \exp\left(i\theta_{10}\right)$. One sees that $A_j = A_{j+Q}$ and $\widetilde{B}_j = \widetilde{B}_{j+Q}$, so that now (7.16) gets implemented as

$$u_j = u_{j+Q} \tag{7.28}$$

This periodicity shows that in order to normalize the wavefunction we have to resort to the discrete integral

$$\sum_{j=1}^{Q} \left|u_j\right|^2 = 1 \quad , \tag{7.29}$$

which is similar to (5.42). Indeed, using the state

$$|\psi\rangle = \sum_{j=1}^{Q} u_j\,(\theta_{10}, \theta_2)\, a_j^+\,(\theta_{10}, \theta_2)\,|0\rangle \tag{7.30}$$

where

$$\left[a_j\left(\theta_{10},\theta_2\right), a_{j'}^{+}\left(\theta_{10},\theta_2\right)\right] = \delta_{j,j'} \tag{7.31}$$

one finds that the $|\psi\rangle$-normalization proceeds as

$$\langle\psi\,|\psi\rangle = \sum_{j=1}^{Q} |u_j|^2 = 1 \tag{7.32}$$

which reproduces precisely (7.29).

It is also of interest to make the substitutions

$$\widetilde{B}_j = \exp\left(ik\right) B_j \tag{7.33}$$

and

$$u_j = \exp\left(-ij\theta_2\right) v_j \tag{7.34}$$

in which case (7.25) becomes

$$B_{j-1}^{*} v_{j-1} + A_j v_j + B_j v_{j+1} = E v_j \quad . \tag{7.35}$$

7.3 The secular equation

Equation (7.35) can be rewritten in a matrix form as

$$\mathfrak{M}_Q \mathcal{U} = E\mathcal{U} \tag{7.36}$$

where $\mathcal{U} = (u_1, u_2, \ldots, u_Q)$ and where $\mathfrak{M}$ is a Hermitian tridiagonal matrix supplemented by non-zero elements at the right top and left bottom:

$$\mathfrak{M}_Q = \begin{pmatrix} A_1 & B_1 \ldots\ldots\ldots & \ldots & B_Q^{*}\exp\left(-iQ\theta_2\right) \\ B_1^{*} & A_2\; B_2 \ldots\ldots & \ldots & \vdots \\ \vdots & B_2^{*}\; A_3\; B_3 \ldots & \ldots & \vdots \\ \vdots & \quad B_3^{*}\; A_4\; B_4 & \ldots & \vdots \\ \vdots & \quad\ddots\;\ddots & \ddots & \vdots \\ \vdots & & A_{Q-1} & B_{Q-1} \\ B_Q\exp\left(iQ\theta_2\right) \ldots & \ldots\ldots & B_{Q-1}^{*} & A_Q \end{pmatrix} \tag{7.37}$$

where $Q > 2$. The energy eigenvalues are given by the roots of the algebraic equation

$$\det\left(\mathfrak{M}_Q - E\right) = 0 \tag{7.38}$$

This amounts to find the roots of a Q-degree polynomial, i.e. to apply the method of the secular equation (Hatsugai and Kohmoto (1990), Wannier et al (1979)). More exactly, (7.38) can be expressed as

$$P_Q\left(E\right) = f\left(\theta_{10}, \theta_2\right) \tag{7.39}$$

for $Q > 2$, in which the polynomial $P_Q\left(E\right)$ is independent of θ_{10} and θ_2, but

$$f\left(\theta_{10}, \theta_2\right) = 2t_a^Q \cos\left(Q\theta_{10}\right) + 2t_b^Q \cos\left(Q\theta_2\right) - 2\left(-1\right)^{P+Q} t_c^Q \cos\left(Q\left(\theta_{10} + \theta_2\right)\right) \tag{7.40}$$

in so far as $t_d = 0$. Of course, the $t_d \neq 0$ -case can also be accounted for, as shown by (2.14) in Hatsugai and Kohmoto (1990). One realizes immediately that (2.24) and (7.39) are quite similar, in the sense that the influence of Brillouin phases is confined exclusively to the r.h.s.'s of corresponding eigenvalue equations. The energy bands are then produced by the inequalities

$$-2t_a^Q - 2t_b^Q \leq P_Q\left(E\right) \leq 2t_a^Q + 2t_b^Q \tag{7.41}$$

in which, once more again, the equality signs are responsible for the band-edges. Inserting $Q = 2$, one realizes that (7.35) exhibits the special form

$$\begin{pmatrix} A_1, & B_1 + B_2^* \exp\left(-2i\theta_2\right) \\ B_1^* + B_2 \exp\left(2i\theta_2\right), & A_2 \end{pmatrix} \begin{pmatrix} v_1 \\ v_2 \end{pmatrix} = E \begin{pmatrix} v_1 \\ v_2 \end{pmatrix} \tag{7.42}$$

where $P = 1$,

$$A_1 = A = -A_2 = 2t_a \cos\theta_{10} \tag{7.43}$$

and

$$C = B_1 + B_2^* \exp\left(-2i\theta_2\right) = \tag{7.44}$$

$$= -2\exp\left(-2i\theta_2\right)\left[t_b \cos\theta_2 - it_c \cos\left(\theta_{10} + \theta_2\right) + it_d \cos\left(\theta_{10} - \theta_2\right)\right] \quad .$$

This gives the energy

$$E = E_{\pm} = \tag{7.45}$$

$$\pm 2\left[t_a^2\cos^2\theta_{10} + t_b^2\cos^2\theta_2 + \left(t_c\cos\left(\theta_{10}+\theta_2\right) - t_d\cos\left(\theta_{10}+\theta_2\right)\right)^2\right]^{1/2}$$

whereas the wavefunction remains to be established by combining the equation

$$v_2 = \frac{C^*}{A+E}v_1 = \frac{E-A}{C}v_1 \tag{7.46}$$

with the normalization condition.

7.4 The $Q = 2$ integral quantum Hall effect

Starting from Kubo's electric transport-formula (Kubo (1966), Datta (1995)), proofs have been given that the single-band contribution to the Hall conductance σ_{xy} is given by (Thouless et al (1982))

$$\sigma_{xy} = \frac{e^2}{2\pi i h}\sum_{l=1}^{2}\oint d\theta_l \widetilde{\mathcal{U}}^* \frac{\partial}{\partial\theta_l}\mathcal{U} \tag{7.47}$$

which exhibits the form of a topological invariant. This formula has been reobtained by accounting for the influence of an additional uniform electric field directed along the y-axis $\overrightarrow{E_F} = (0, E_F, 0)$, now by invoking the gauge $\overrightarrow{E}_F = -\partial\overrightarrow{A}/\partial t$ (Kohmoto (1989)). In this latter case the tight-binding description works in accord with a modified form of (7.11). Indeed, the θ_{ji}-phases discussed before undergo well-defined modifications under the influence of the electric field like

$$\theta_{ji} = 2\pi\phi_n \rightarrow 2\pi\phi_n - E_F t \tag{7.48}$$

where $i = (n, m)$ and $j = (n, m+1)$ and similarly for other cases. The contour in (7.47) is a rectangle encircling Brillouin zones for which $-\pi/Q \leq \theta_{10} \leq \pi/Q$ and $0 \leq \theta_2 \leq 2\pi$. Keeping in mind (7.19), we shall use by now the more convenient $\theta_1 \equiv \theta_{10}$-quotation. Introducing the phases in terms of the factorization

$$u_j\left(\vec{\theta}\right) = u_j\left(\theta_1, \theta_2\right) = \left|u_j\left(\vec{\theta}\right)\right| \exp\left(i\beta_j\left(\vec{\theta}\right)\right) \tag{7.49}$$

one sees that (7.47) becomes

$$\sigma_{xy} = \frac{e^2}{2\pi h}\sum_{j=1}^{Q}\oint d\vec{\theta}\cdot\frac{\partial}{\partial\vec{\theta}}\beta_j\left(\vec{\theta}\right)\left|u_j\left(\vec{\theta}\right)\right|^2 \quad . \tag{7.50}$$

What then remains is to establish the wavefunction in the wave-vector representation.

Coming back to the previous section let us assume that u_1 is real, such that

$$u_1\left(\vec{\theta}\right) = u_1^*\left(\vec{\theta}\right) \equiv -\sin\alpha \tag{7.51}$$

where, for the moment, the α-angle is arbitrary. Choosing e.g. $E = E_-$, we then find that

$$u_2\left(\vec{\theta}\right) = \exp\left(i\zeta\right)\cos\alpha \tag{7.52}$$

in accord with (7.46), so that

$$\zeta = \zeta_1\left(\vec{\theta}\right) = -\arctan\left(\frac{t_c + t_d}{t_b}\tan\left(\theta_2\right)\right) \tag{7.53}$$

for $\theta_1 = \pi/2$, but $\zeta = -\zeta_1\left(\vec{\theta}\right)$, if $\theta_1 = -\pi/2$. In addition, we have to consider that $\tan^2\alpha = 1$, as soon as $\theta_1 = \pm\pi/2$. Under such circumstances (7.50) becomes

$$\sigma_{xy} = \frac{e^2}{2\pi h}\oint\left|u_2\right|^2 d\zeta \tag{7.54}$$

which can be easily evaluated with the help of integral (see (2.562.1) in Gradsteyn and Ryzhik (1965)):

$$\int\frac{dx}{b^2 + c^2\sin^2\left(ax\right)} = \frac{1}{ab\sqrt{b^2+c^2}}\arctan\frac{\sqrt{b^2+c^2}\tan\left(ax\right)}{b} \quad . \tag{7.55}$$

Proceeding in this manner leads to

$$\sigma_{xy} = \frac{e^2}{h} n_H \tag{7.56}$$

which stands for the quantized Hall conductivity, now for $n_H = 1$. In general, there is $n_H = 1, 2, 3, \ldots$, which can be readily explained by virtue of topological properties (Hatsugai and Kohmoto (1990)). However, experimental evidence has been found not only for the integral quantum Hall-effect (Czerwinski and Brown (1991), von Klitzing et al (1980)), but also for the fractional one for which $n = 1/3, 2/3, 2/7, 2/5, \ldots$ (Tsui et al (1982), Ishikawa and Maeda (1997)). Needless to say that such problems are still highly interesting nowadays (Fradkin (1991), Halperin (1984), Bellisard et al (1994), Huckenstein (1995)).

7.5 Duality properties

Let us make the discrete Fourier transformation

$$u_j \to f_l = \sum_{l=1}^{Q} \exp(2i\pi\beta jl)\, u_j \tag{7.57}$$

which can be easily inverted as

$$u_j = \frac{1}{Q} \sum_{l=1}^{Q} \exp(-2i\pi\beta jl)\, f_l \tag{7.58}$$

by virtue of the orthogonality condition

$$\sum_{l=1}^{Q} \exp 2i\pi\beta\,(j - j\prime)\, l = Q\delta_{j',j} \quad . \tag{7.59}$$

Then (7.25) gets reproduced under (7.58) in terms of the mappings

$$u_j \to f_j \tag{7.60}$$

$$(\theta_1, \theta_2) \to (\theta_2, \theta_1) \tag{7.61}$$

$$\beta \to -\beta \tag{7.62}$$

and

$$(t_a, t_b, t_c, t_d) \to (t_b, t_a, t_c, t_d) \tag{7.63}$$

where $\theta_{10} \equiv \theta_1$, as already noted before. Equation (7.62) can be rewritten equivalently as $q \to 1/q$, which amounts to reverse the direction of the magnetic field. Comparing (7.60)-(7.63) with (7.39), we have to realize that the energy polynomial $P_Q(E)$ is itself invariant under above duality transformations.

7.6 Tight binding descriptions with inter-band couplings

The general form of the 2D periodic potential reads

$$V(x_1, x_2) = V_0 \sum_{r,s} v_{r,s} \exp 2\pi i \left(r\frac{x_1}{a} + s\frac{x_2}{b} \right) \tag{7.64}$$

so that

$$V(x_1 + a, x_2 + b) = V(x_1, x_2) \quad . \tag{7.65}$$

Under the influence of a transversal and homogeneous magnetic field $\overrightarrow{B} = (0, 0, B)$, the electron is described by the Schrödinger-equation

$$\left[\frac{1}{2m} \left(-i\hbar \nabla_2 + \frac{e}{c}\overrightarrow{A} \right)^2 + V(x_1, x_2) \right] \psi(x_1, x_2) = E\psi(x_1, x_2) \tag{7.66}$$

where $m = m^*$ stands for the effective mass, as usual. Choosing the Landau gauge $\overrightarrow{A} = B(0, x_1, 0)$, one works in conjunction with the factorization

$$\psi(x_1, x_2) = \exp(ik_2 x_2)\, \psi_{oscill}(x_1) \quad . \tag{7.67}$$

A similar equation will be invoked later in the derivation of the Harper-equation. Ignoring, for the moment, the influence of the periodic potential leads to the Landau levels

$$E_\nu = \hbar\omega_c \left(\nu + \frac{1}{2} \right) \tag{7.68}$$

where $\omega_c = eB/mc$ is the cyclotron frequency, while $\nu = 0, 1, 2, \ldots$. This proceeds in terms of the oscillator wave function

$$\psi_{\nu,\theta}\left(x_1,x_2\right)=\left\langle x_1,x_2\right|\nu,\theta\rangle=N_0\exp\left(i\frac{x_2}{b}\theta\right)\psi_\nu\left(z\right)\tag{7.69}$$

where $\theta\equiv\theta_2=bk_2$. One has

$$\psi_{oscill}\left(x_1\right)=\psi_\nu\left(z\right)=\exp\left(-z^2/2\right)H_\nu\left(z\right)\tag{7.70}$$

where $H_\nu\left(z\right)$ denotes the Hermite-polynomial. The related dimensionless coordinate is given by

$$z=\frac{x_1}{l_B}+\theta\frac{l_B}{b}\tag{7.71}$$

in which l_B is the magnetic length introduced before via (2.77). This wave-function is normalized as

$$\left\langle\nu',\theta'\right|\nu,\theta\rangle=\delta_{\nu,\nu'}\delta\left(\theta-\theta'\right)\tag{7.72}$$

so that

$$N_0=\frac{1}{\left[\pi^{3/2}bl_B2^{\nu+1}\Gamma\left(\nu+1\right)\right]^{1/2}}\quad.\tag{7.73}$$

The next task is the derivation of the matrix elements of the periodic potential. After some lengthy calculations, the following result has been written down (Petschel and Geisel (1993), Springsguth et al (1997)):

$$\nu_{r,s}\left\langle\nu',\theta'\right|\exp 2\pi i\left(r\frac{x_1}{a}+s\frac{x_2}{b}\right)\left|\nu,\theta\right\rangle=\tag{7.74}$$

$$=P_{\nu'\nu}\left(r,s\right)\exp\left(-i\frac{\phi_0}{\phi}r\theta'\right)\delta\left(\theta'-\theta-2\pi s\right)\quad.$$

One has

$$P_{\nu'\nu}\left(r,s\right)=P\left(\nu',\nu;r,s\right)=\nu_{r,s}\exp\left(irs\pi\frac{\phi_0}{\phi}\right)\exp\left(-\frac{u}{2}\right)\cdot\tag{7.75}$$

$$\cdot\sqrt{\frac{\nu!}{\nu'!}}\left(\pi\frac{\phi_0}{\phi}\right)^{\left(\nu'-\nu\right)/2}\frac{L_\nu^{\left(\nu'-\nu\right)}\left(u\right)}{\left(\pm s\sqrt{\frac{a}{b}}+ir\sqrt{\frac{b}{a}}\right)^{\nu-\nu'}}$$

when $\nu' \geqslant \nu$, where $L_\nu^{(\alpha)}(u)$ is the Laguerre polynomial and

$$u = \pi\frac{\phi_0}{\phi}\left(r^2\frac{b^2}{a^2} + s^2\frac{a^2}{b^2}\right) \quad . \tag{7.76}$$

Performing the substitutions $\nu' \rightleftarrows \nu$ yields

$$P_{\nu'\nu}(r,s) = P(\nu,\nu';r,s) \tag{7.77}$$

instead of (7.75), which is valid this time when $\nu \geqslant \nu'$. In addition, $+s(-s)$ stands for $\nu' \geqslant \nu(\nu \geqslant \nu')$. Such matrix elements have also been discussed before (see e.g. (7) in Pfannkuche and Gerhardts (1992) or (7) and (8) in Springsguth et al (1997)). Accordingly, the matrix elements of the Hamiltonian are given by

$$\langle\nu',\theta'|\,\mathcal{H}\,|\nu,\theta\rangle = \hbar\omega_c\left(\nu+\frac{1}{2}\right)\delta_{\nu,\nu'}\delta(\theta-\theta')+ \tag{7.78}$$

$$+V_0\sum_{r,s}P_{\nu'\nu}(r,s)\exp\left(-i\frac{\phi_0}{\phi}r(\theta+2\pi s)\right)\delta(\theta'-\theta-2\pi s)$$

which serves as a starting point towards further developments. To this aim let us perform again the extra θ-discretization

$$\theta = 2\pi\widetilde{n} + \widetilde{\theta} \tag{7.79}$$

which is reminiscent to (7.19), where $\widetilde{n}$ is an integer and $\widetilde{\theta} \in [0, 2\pi)$. Correspondingly, the orthogonality equation (7.72) becomes

$$\left\langle\nu',\widetilde{n}',\widetilde{\theta}'\middle|\,\nu,\widetilde{n},\widetilde{\theta}\right\rangle = \delta_{\nu,\nu'}\delta_{\widetilde{n},\widetilde{n}'}\delta\left(\widetilde{\theta}'-\widetilde{\theta}\right) \quad . \tag{7.80}$$

On the other hand, the energy eigenfunction can be decomposed as

$$\left|E;\widetilde{\theta}\right\rangle = \sum_{\nu,\widetilde{n}} a_{\widetilde{n}}^{\nu}\left(\widetilde{\theta}\right)\left|\nu,\widetilde{n},\widetilde{\theta}\right\rangle \tag{7.81}$$

such that

$$\mathcal{H}\left|E;\widetilde{\theta}\right\rangle = E\left|E;\widetilde{\theta}\right\rangle \tag{7.82}$$

where E plays the role of the band energy. We have also to remark that the matrix elements of the Hamiltonian are diagonal in $\widetilde{\theta}$ since

$$\delta\left(\theta' - \theta - 2\pi s\right) = \delta_{\widetilde{n}', \widetilde{n}+s}\delta\left(\widetilde{\theta}' - \widetilde{\theta}\right) \tag{7.83}$$

by virtue of (7.79).

Under such conditions (7.82) produces the tight binding equation

$$\sum_{\nu} a_{\widetilde{n}'}^{\nu}\left(\widetilde{\theta}'\right) B_{\nu}^{\nu'}\left(\widetilde{n}', \widetilde{\theta}'\right) + \sum_{\nu, s\neq 0} a_{\widetilde{n}'-s}^{\nu}\left(\widetilde{\theta}'\right) C_{\nu}^{\nu'}\left(s, \widetilde{n}', \widetilde{\theta}'\right) = \tag{7.84}$$

$$= \left[E - \hbar\omega_c\left(\nu' + \frac{1}{2}\right)\right] a_{\widetilde{n}'}^{\nu'}\left(\widetilde{\theta}'\right)$$

where

$$B_{\nu}^{\nu'}\left(\widetilde{n}', \widetilde{\theta}'\right) = V_0 \sum_{r} P_{\nu',\nu}\left(r, 0\right) \exp\left(-i\frac{\phi_0}{\phi} r\left(\widetilde{\theta}' + 2\pi\widetilde{n}'\right)\right) \tag{7.85}$$

and

$$C_{\nu}^{\nu'}\left(s, \widetilde{n}', \widetilde{\theta}'\right) = V_0 \sum_{r} P_{\nu',\nu}\left(r, s\right) \exp\left(-i\frac{\phi_0}{\phi} r\left(\widetilde{\theta}' + 2\pi\widetilde{n}'\right)\right) \quad . \tag{7.86}$$

One recognizes that (7.84) is able to account for inter-band couplings if there are non-zero realizations of $B_{\nu}^{\nu'}$ and/or $C_{\nu}^{\nu'}$ for $\nu' \neq \nu$. The strength of such couplings is proportional to the quotient $V_0/\hbar\omega_c$. Making the identification

$$\frac{V_0}{\hbar\omega_c} = K\frac{\phi_0}{\phi} \tag{7.87}$$

yields $K = 2\pi mabV_0/h^2$, which is independent of the number of flux quanta per unit cell and which provides a useful measure of the Landau inter-band coupling. Hopping effects can also be identified in terms of s-values for which $\nu_{r,s} \neq 0$. Equation (7.84) has been handled numerically, which results in interesting nested structures concerning both energy-bands and the Hall conductance (Springsguth et al (1997)). Disorder broadening effects have also been discussed before (Ando et al (1982)).

7.7 Concrete single-band equations and classical realizations

Next let us preserve only the lowest Fourier components in (7.64). This yields the periodic potential

$$V(x_1, x_2) = V_0 \left(\cos\left(2\pi \frac{x_1}{a}\right) + \cos\left(2\pi \frac{x_2}{a}\right)\right) \tag{7.88}$$

in which this time $a = b$ and

$$v_{1,0} = v_{-1,0} = v_{0,1} = v_{0,-1} = \frac{1}{2} \quad . \tag{7.89}$$

Ruling out the inter-band couplings leads to the alternative NN equation

$$a^{\nu}_{\tilde{n}-1} + a^{\nu}_{\tilde{n}+1} + 2\cos\frac{\phi_0}{\phi}\left(2\pi\tilde{n} + \tilde{\theta}\right) a^{\nu}_{\tilde{n}} = \mathcal{E}_s a^{\nu}_{\tilde{n}} \quad . \tag{7.90}$$

The scaled energy is given by

$$\mathcal{E}_s = \frac{E - \hbar\omega_c\left(\nu + \frac{1}{2}\right)}{V_0 P_{\nu\nu}(1,0)} \tag{7.91}$$

where

$$P_{\nu\nu}(\pm 1, 0) = P_{\nu\nu}(0, \pm 1) = \frac{1}{2}\exp\left(-\frac{\pi}{2}\frac{\phi_0}{\phi}\right) L^{(0)}_{\nu}\left(\pi\frac{\phi_0}{\phi}\right) \quad . \tag{7.92}$$

The zeros of the Laguerre-polynomial $L^{(0)}_{\nu}(\pi\phi_0/\phi)$ displayed above, say $\pi\phi_0/\phi = z_0$, have to be considered, too. Indeed (7.91) indicates that one has $E = E_\nu$ when

$$\phi = \phi_\nu = \pi\phi_0/z_0 \tag{7.93}$$

and $\nu \neq 0$, which means that the Landau energy-band becomes a zero-width energy level.

One remarks that (7.90), which can be traced back to the paper by Rauh (1975), looks like the Harper-equation (8.7). However, the commensurability parameter ϕ/ϕ_0 is implemented this time in a reversed manner. In addition, the phase characterizing (7.90) for $\tilde{n} = 0$ is sensitive to the magnetic flux, which differs from the $n = 0$ form of (8.7). So far we then have to realize that (7.90) and (8.7) rely on complementary degrees of validity, i.e.

on the complementary limits of large and small B-values, respectively. The influence of higher-order neighbors can be established in a similar manner by accounting for higher-harmonics in (7.64).

Shifting the coordinate, one realizes that the classical counterpart of the quantum-mechanical Hamiltonian characterizing (7.90) is given by

$$\mathcal{H}_{cl}^{(LG)} = 2\cos(p) + 2\cos(2\pi\beta n) \tag{7.94}$$

where p originates from the dimensionless momentum operator $-i\partial/\partial n$. This Hamiltonian is integrable, as it produces just regular trajectories in the Poincaré-section. On the other hand, chaotic trajectories are useful in the description of lateral surface supperlattices. To this aim a classical Hamiltonian producing the chaotic behavior needed can be easily derived. Indeed, let us insert the symmetric gauge $\overrightarrow{A} = (\overrightarrow{B} \times \overrightarrow{x})/2$ into the classical Hamiltonian

$$\mathcal{H}_{cl} = \frac{1}{2m}\left(\overrightarrow{p} + \frac{e}{c}\overrightarrow{A}\right)^2 + V(x_1, x_2) \tag{7.95}$$

concerning (7.66). Introducing new dimensionless phase-space coordinates X_j and P_j ($j = 1, 2$) via (Petschel and Geisel (1993))

$$X_1 = \frac{\sqrt{ab}}{\gamma}\left(p_1 + \frac{e}{2c}Bx_2\right), \quad P_1 = \frac{\sqrt{ab}}{\gamma}\left(p_2 - \frac{e}{2c}Bx_1\right) \tag{7.96}$$

and

$$X_2 = \frac{\sqrt{ab}}{\gamma}\left(-p_1 + \frac{e}{2c}Bx_2\right), \quad P_2 = \frac{\sqrt{ab}}{\gamma}\left(p_2 + \frac{e}{2c}Bx_1\right) \tag{7.97}$$

yields the converted Hamiltonian

$$\mathcal{H}_{cl}^{(SG)}(\overrightarrow{X}, \overrightarrow{P}) = \frac{\Omega}{2}\left(X_1^2 + P_1^2\right) + V_0\left[\cos(P_2 - P_1) + \cos(X_1 + X_2)\right] \tag{7.98}$$

in accord with the $a \neq b$ version of (7.88), where

$$\Omega = \frac{eB}{mc}\frac{\gamma}{2\pi} \tag{7.99}$$

and

$$\gamma = \frac{eBab}{2\pi c} = \hbar \frac{\Phi}{\Phi_0} \quad . \tag{7.100}$$

This Hamiltonian leads in turn to chaotic trajectories for small but non-zero values of the quotient Ω/V_0, as indicated by the Poincaré-sections displayed in Fig. 1 in Petschel and Geisel (1993).

It has also been found, that the electrons exhibit ballistic transport (localization) in the direction of strong potential (weaker) modulation (Ketzmerick et al (2000)). More exactly, choosing the 2D periodic potential

$$V(x_1, x_2) = V_1 \cos\left(2\pi \frac{x_1}{a}\right) + V_2 \cos\left(2\pi \frac{x_2}{a}\right) \tag{7.101}$$

instead of (7.88), one would then have to consider the x_1- and x_2-directions, respectively, where now $V_1 > V_2 > 0$ instead of $V_1 = V_2$. Equations (7.66) and (7.101) are also responsible for the appearance of avoided band crossings, which leads again to chaotic dynamics in the classical limit, as one might expect.

Chapter 8

The Harper-Equation and Electrons on the $1D$ Ring

The celebrated equation written down by Harper (1955) is still a quite fascinating problem producing rich structures and unexpected implementations since half a century (Rauh (1975), Wannier (1975), Obermair and Wannier (1976), Hofstadter (1976), Sokoloff (1985), Hiramoto and Kohmoto (1989) Gredeskul et al (1997)). We have to specify that this equation serves to the description of a special nanosized system , i.e. of the two dimensional electron gas ($2DEG$) on a $2D$ square lattice with NN-hoppings under the influence of a transversal and homogeneous magnetic field $\overrightarrow{B} = (0, 0, B)$. Note that the 2DEG has been produced by virtue of modulation-doping techniques, such as realized in GaAs/(Ga,Al)As -heterostructures (Bastard (1992)). The same equation concerns the description of a 2D lattice of quantum dots under the influence of the magnetic field.

The interesting point is that the Harper-equation lies at the confluence of several research fields, such as superconductivity proceeding in terms of linearized Ginzburg-Landau equations (Wang et al (1987)), the d-wave superconductivity with a magnetic field (Morita and Hatsugai (2001)), level statistics in quantum systems with unbounded diffusion (Geisel et al (1991)), critical quantum chaos (Evangelou and Pichard (2000)), or anomalous diffusion of wave packets in quasi-periodic chains (Piéchon (1996)). In addition, there are several related problems already referred to before such as localization length and metal-insulator transition (Aubry and Andre (1980), Sokoloff (1981), Thouless (1983a)), the quantum Hall-effect (Thouless et al (1982), Czerwinsky and Brown (1991), von Klitzing et al (1980), Tsui et al (1982), Ishikava and Maeda (1997), Halperin (1984), Bellisard et al (1994), Huckenstein (1995), Springsguth et al (1997)), and last but not least interactions in Aharonov-Bohm cages (Vidal et al (1998)).

The influence of the magnetic field on electrons moving on the 1D ring has also received much interest. This serves as a useful model to the theoretical description of nanorings. We shall then present the derivation of total persistent currents at $T = 0$ by proceeding in a close connection with the methods presented before by Cheung et al (1988).

8.1 The usual derivation of the Harper-equation

The Harper-equation originates from the influence of the minimal substitution on a $2D$ energy-dispersion law

$$\mathcal{E}_d\left(\overrightarrow{\theta}\right) = \cos\theta_1 + \Delta\cos\theta_2 \tag{8.1}$$

where Δ is an anisotropy parameter and where, as before, the Brillouin phases are denoted by $\theta_l = k_l a$ $(l = 1, 2)$. Of course, (8.1) can be viewed as a $2D$ generalization of (6.4). Choosing the Landau-gauge

$$\overrightarrow{A} = (0, Bx_1, 0) \tag{8.2}$$

and applying the minimal substitution in terms of the wavevector operator

$$\overrightarrow{k}_{op} = \frac{1}{\hbar}\overrightarrow{p}_{op} = -i\nabla \tag{8.3}$$

yields the substitution rule

$$k_l \rightarrow k_l^{(op)} = -i\frac{\partial}{\partial x_l} + \frac{e}{\hbar c}A_l \quad . \tag{8.4}$$

Combining (8.1) and (8.4) generates the $2D$ Hamiltonian (Harper (1955), Wilkinson (1986))

$$H = \cos\left(-ia\frac{\partial}{\partial x_1}\right) + \Delta\cos\left(a\left(-i\frac{\partial}{\partial x_2} + \frac{e}{\hbar c}Bx_1\right)\right) \tag{8.5}$$

which can also be interpreted in terms of magnetic translations , as we shall see later. The application of the minimal substitution (8.4) to (8.1) can be traced back to Peierls (1933). Next we shall factorize the wavefuntion as

$$\psi(x_1, x_2) = \exp(ik_2x_2)\,\psi(x_1) \quad , \tag{8.6}$$

which opens the way to the $1D$ reduction. Indeed, performing the translations and accounting for (7.5) results in the celebrated Harper-equation , i.e. in the $1D$ second-order discrete equation

$$\varphi_{n+1} + \varphi_{n-1} + 2\Delta\cos(2\pi\beta n + \theta_2)\,\varphi_n = E\varphi_n \tag{8.7}$$

where $\varphi_n = \psi(na)$, $x_1 = na$ and $E = 2\mathcal{E}_d$. One sees that (8.7) is a particular realization of (7.35) discussed before. However, (8.7) exhibits rather interesting implementations, so that it has its own theoretical interest. We can also resort to the factorization

$$\varphi_n = \exp(in\theta_1)\,u_n \tag{8.8}$$

in which case (8.7) is replaced by

$$\exp(i\theta_1)\,u_{n+1} + \exp(-i\theta_1)\,u_{n-1} + 2\Delta\cos(2\pi\beta n + \theta_2)\,u_n = Eu_n \tag{8.9}$$

which incorporates, this time, both Brillouin phases.

8.2 The transfer matrix

Considering three successive sites, one realizes that (8.7) can be rewritten as

$$\begin{pmatrix}\varphi_{n+1}\\ \varphi_n\end{pmatrix} = \mathbb{T}_n\begin{pmatrix}\varphi_n\\ \varphi_{n-1}\end{pmatrix} \equiv \begin{pmatrix}E - 2\Delta\cos(2\pi\beta n + \theta_2) & , -1\\ 1 & , \ 0\end{pmatrix}\begin{pmatrix}\varphi_n\\ \varphi_{n-1}\end{pmatrix} \tag{8.10}$$

where by construction

$$\det\mathbb{T}_n = 1 \quad . \tag{8.11}$$

Starting from $n = 0$ and performing successive steps, yields the relationship

$$\begin{pmatrix}\varphi_Q\\ \varphi_{Q-1}\end{pmatrix} = \mathcal{T}_Q\begin{pmatrix}\varphi_0\\ \varphi_{-1}\end{pmatrix} = \prod_{n=0}^{Q-1}\mathbb{T}_n\begin{pmatrix}\varphi_0\\ \varphi_{-1}\end{pmatrix} \tag{8.12}$$

in which $\mathcal{T}_Q$ stands for the so called transfer matrix (Hofstadter (1976), Kohmoto et al (1983)) It is clear that this matrix obeys the equation

$$\det \mathcal{T}_Q = 1 \tag{8.13}$$

by virtue of (8.11). On the other hand, one has the periodicity condition

$$\varphi_{n+Q} = \varphi_n \tag{8.14}$$

in accord with (7.4), so that

$$|\varphi_Q|^2 + |\varphi_{Q-1}|^2 = |\varphi_0|^2 + |\varphi_{-1}|^2 \quad . \tag{8.15}$$

In other words the $\mathcal{T}$ -matrix is also unitary, so that the corresponding eigenvalues exhibit the general form

$$\lambda_\pm = \exp(\pm i\mu_1) \tag{8.16}$$

where μ_1 is a real parameter. We then get immediately the trace formula (Hofstadter (1976))

$$Tr\mathcal{T}_Q = 2\cos\mu_1 \tag{8.17}$$

which relies on (7.39). Indeed, one has

$$Tr\mathcal{T}_Q = P_Q(E;\Delta) - 2\Delta^Q \cos(Q\theta_2) \tag{8.18}$$

which meets (7.39) via $t_a = 1$, $t_b = \Delta$, $Q\theta_1 = \mu_1$ and $t_c = 0$.

Proofs have also been given that there is a limiting matrix like (Huckenstein (1995), Deych et al (2003))

$$\Gamma_L = \lim_{Q\to\infty} (\mathcal{T}_Q^* \mathcal{T}_Q)^{1/2Q} \quad . \tag{8.19}$$

The Lyapunov exponents characterizing $\mathcal{T}_Q$, i.e. γ_1 and γ_2, are exhibited by the eigenvalues $\exp(\gamma_1)$ and $\exp(\gamma_2)$ of this matrix. The inverse localization length is then given by the Lyapunov exponent of smallest absolute magnitude.

8.3 The derivation of Δ-dependent energy polynomials

Explicit evaluations of energy polynomials are of interest for further applications. Starting from $\beta = P/Q$ and fixing the Q-denominator yields an even number, say $2\mathcal{N}_s(Q)$, of coprime $P \equiv P_s$-realizations of the P-nominator. This in turn produces $\mathcal{N}_s(Q)$ distinct realizations of the energy-polynomial like $P_Q^{(k)}(E;\Delta)$, where $k = 1, 2, \ldots, \mathcal{N}_s(Q)$. One has e.g.

$$\mathcal{N}_s(1) = \mathcal{N}_s(2) = \mathcal{N}_s(3) = \mathcal{N}_s(4) = \mathcal{N}_s(6) = 1 \tag{8.20}$$

and $\mathcal{N}_s(5) = \mathcal{N}_s(8) = 2$, but $\mathcal{N}_s(7) = 3$ and so one. Of course, if $\mathcal{N}_s(Q) = 1$ one has $P = 1$ and $P = Q - 1$, only. It is further clear, that the energy polynomial $P_Q(E;\Delta)$ exhibits Q real roots labelled as

$$E = E_{j,k}^{(Q)}(\mu_1, \mu_2; b) \tag{8.21}$$

where $b = \Delta^Q$, $j = 1, 2, \ldots, Q$ and $\mu_l = Q\theta_l$. So it is enough to assume that

$$\mu_l \in [0, \pi] \tag{8.22}$$

which produces a number of Q bands via

$$-2 - 2b \leq P_Q^{(k)}(E;\Delta) \leq 2 + 2b \tag{8.23}$$

in accord with (7.41).

The polynomials for $Q = 1 - 8$ are as follows (Hatsugai and Kohmoto (1990), Papp et al (2002a), Hong and Salk (1999))

$$P_1 = E \qquad \text{and} \qquad P_2 = E^2 - 2\left(\Delta^2 + 1\right) \tag{8.24}$$

whereas for $Q \geq 3$ there is

$$P_3 = E\left(E^2 - 3\left(\Delta^2 + 1\right)\right) \quad , \tag{8.25}$$

$$P_4 = E^4 - 4E^2\left(\Delta^2 + 1\right) + 2\left(\Delta^4 + 1\right) \tag{8.26}$$

$$P_5^{(\pm)} = E^5 - 5E^3\left(\Delta^2 + 1\right) + 5E\left(\left(\Delta^4 + 1\right) + \frac{1}{2}\Delta^2\left(3 \pm \sqrt{5}\right)\right) \tag{8.27}$$

$$P_6 = E^6 - 6E^4\left(1+\Delta^2\right) + 3E^2\left(3+2\Delta^2+3\Delta^4\right) - 2\left(\Delta^6+1\right) \quad (8.28)$$

$$P_7^{(k)} = \quad (8.29)$$

$$= E\left(E^6 - 7E^4\left(1+\Delta^2\right) + 14E^2\left(1+\Delta^4\right)\right.$$

$$\left. + \widetilde{s}_k E^2\Delta^2 - \widetilde{t}_k\Delta^2\left(1+\Delta^2\right) - 7\left(1+\Delta^6\right)\right)$$

where now $k = 1, 2, 3$ and

$$P_8^{(k)}(E,\Delta) = \quad (8.30)$$

$$= E^8 - 8E^6\left(1+\Delta^2\right) + 20E^4\left(1+\Delta^4\right) + +8\left(4\pm\sqrt{2}\right)E^4\Delta^2 -$$

$$-16E^2\left(1+\Delta^6\right) - 16\left(2\pm\sqrt{2}\right)E^2\Delta^2\left(1+\Delta^2\right) + 2\left(1+\Delta^8\right) \quad .$$

The $\widetilde{s}_k$ and $\widetilde{t}_k$-parameters in (8.29) and (8.30) are given by

$$\widetilde{s}_1 = 7\left(3 - 2\cos\frac{2\pi}{7}\right) \cong 12.271143 \quad (8.31)$$

$$\widetilde{s}_2 = 7\left(3 + 2\cos\frac{3\pi}{7}\right) \cong 24.115293 \quad (8.32)$$

and

$$\widetilde{t}_1 = 14 - 14\cos\frac{2\pi}{7} - 14\cos\frac{3\pi}{7} \cong 2.155849 \quad (8.33)$$

$$\widetilde{t}_2 = 7 - 14\cos\frac{2\pi}{7} + 28\cos\frac{3\pi}{7} \cong 4.501739 \quad (8.34)$$

and

$$\widetilde{t}_3 = 21 + 28\cos\frac{2\pi}{7} - 14\cos\frac{3\pi}{7} \cong 35.342421 \quad (8.35)$$

respectively. We have to note that the $Q = 3$-case has received much attention (Wang et al (1987), Lipan (2000)). For larger Q-values we have to resort, of course, to numerical estimates (Hofstadter (1976), Wannier

(1978)). Performing unions over Q's produces interesting self-similar nested band-structures, referred to as the Hofstadter butterfly (Hofstadter (1976)), such as displayed by the $\beta - E$-diagram in figure 8.1 for $Q < 50$, $\theta_2 = 0$ and $\Delta = 1$. So far the butterfly type spectrum has received experimental supports within the high field regime (Gerhardts et al (1991)). Moreover, the Hofstadter butterfly referred to above can also be understood as a phase diagram of the integer quantum Hall effect (Klitzing et al (1980), Klitzing (1986), Albrecht et al (2001)). This time one has infinitely many phases which are discriminated by virtue of Chern-numbers characterizing quantized Hall-conductances (Osadchy and Avron (2001)).

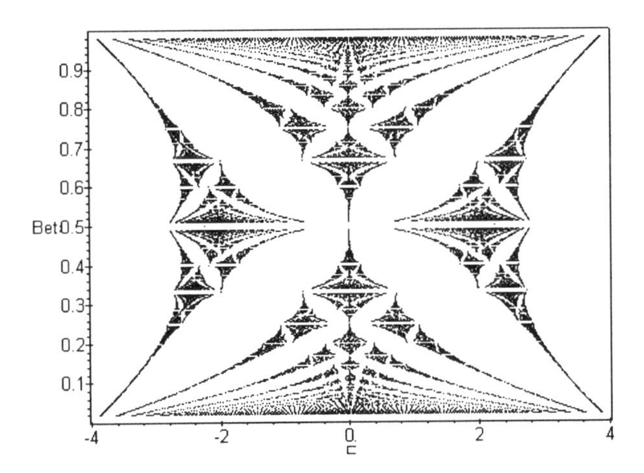

Fig. 8.1 **The nested band structure characterizing the Harper-equation (7.7) for $Q < 50$, $\theta_2 = 0$, $\Delta = 1$ and $\beta \in [[0, 1]]$.**

8.4 Deriving Δ-dependent DOS-evaluations

The above energy-polynomials can be readily applied in order to obtain updated DOS -evaluations. One starts by expressing the average value of a function $A = A(E; b, \mu_1, \mu_2)$ as

$$\langle A \rangle = \frac{1}{\pi^2 Q \; N_s\,(Q)} \sum_{j=1}^{Q} \sum_{k=1}^{N_s} I(b) \tag{8.36}$$

where (Wannier et al (1979))

$$I(b) = \int_0^{\pi} d\mu_1 \int_0^{\pi} d\mu_2 A(E) \quad . \tag{8.37}$$

Using for instance the shorthand quotation $P = P_Q^{(k)}(E;\Delta)$ one obtains

$$d\mu_2 = -f\,(\mu_1, P)\,\frac{dP}{2b} = -\left[1 - \frac{1}{b^2}\left(\frac{P}{2} - \cos\mu_1\right)^2\right]^{-1/2} \frac{dP}{2b} \tag{8.38}$$

in accord with (8.17) and (8.18), which also means that $dP = |(dP/dE)|\, dE$. Keeping in mind that we are interested in the derivation of positive density contributions, we can organize the integral in (8.37) as

$$I(b) = +\frac{1}{2b}\left[\int_{-2+2b}^{2+2b} dP\;A(E)F_1 + \int_{-2-2b}^{2-2b} dP\;A(E)F_2 + \int_{2-2b}^{-2+2b} dP\;A(E)F_3\right] \tag{8.39}$$

where

$$F_1 = F_1(P^2, b) = \int_{\arccos(P/2+b)}^{\pi} d\mu_1 f\,(\mu_1, P) = \frac{4b}{\sqrt{-U_-}} K(S) \tag{8.40}$$

$$F_2 = F_2(P^2, b) = \int_0^{\arccos(P/2-b)} d\mu_1 f\,(\mu_1, P) = \sqrt{b} K\left(\frac{1}{4}\sqrt{\frac{U_+}{b}}\right) \tag{8.41}$$

and

$$F_3 = F_3(P^2, b) = \int_{\arccos(P/2+b)}^{\arccos(P/2-b)} d\mu_1 f\,(\mu_1, P) = \frac{4b}{\sqrt{U_+}} K\left(4\sqrt{\frac{b}{U_+}}\right) \tag{8.42}$$

where K and F are the complete and incomplete elliptic integrals of the first kind, respectively. One has $U_\pm = (2b \pm 2)^2 - P^2$ and $S = \sqrt{U_+/U_-}$. It can also be easily verified that

$$\frac{4b}{\sqrt{-U_-}} K(S) = \sqrt{b} K\left(\frac{1}{4}\sqrt{\frac{U_+}{b}}\right) \tag{8.43}$$

as shown by (15.34) and (15.35) in Abramowitz and Stegun (1972), which also means that $F_1(P^2, b) = F_2(P^2, b)$. Symmetry properties like

$$\frac{1}{\sqrt{U_-}} K\left(4\sqrt{\frac{b}{-U_-}}\right) = \frac{1}{\sqrt{U_+}} K\left(4\sqrt{\frac{b}{U_+}}\right) \tag{8.44}$$

or

$$F_2(P^2, -b) = \frac{4b}{\sqrt{-U_-}} F\left(\frac{1}{S}, S\right) \tag{8.45}$$

where $F(1/S, S) = K(1/S)/S$, can also be mentioned. Next one realizes that

$$\widetilde{F}_0(P^2, b) \equiv Re F_1(P^2, b) = Re F_3(P^2, b) = \sqrt{b} Re K'\left(\frac{1}{4}\sqrt{\frac{-U_-}{b}}\right) > 0 \tag{8.46}$$

where K' is the associated complete elliptic integral of the first kind. This produces an apparent generalization of the $b = 1$ -result written down long ago (Wannier et al (1979)). Putting together above results then gives the DOS as

$$D_K^{(Q)}(E; \Delta) = \sum_{k=1}^{\mathcal{N}_s(Q)} g_k^{(Q)}(E, b) \tag{8.47}$$

where

$$g_k^{(Q)}(E, b) = \frac{1}{2\pi^2 b Q N_s(Q)} \left| \frac{dP_Q^{(k)}(E; \Delta)}{dE} \right| \widetilde{F}\left(\left(P_Q^{(k)}(E; \Delta)\right)^2, b\right) \tag{8.48}$$

and

$$\widetilde{F}\left(P^2, b\right) = \widetilde{F}_0(P^2, b)\, \theta(U_+) \tag{8.49}$$

in which θ denotes the Heaviside function. This DOS gets normalized as

$$\int_{-2-2\Delta}^{2+2\Delta} D_K^{(Q)}(E;\Delta)\,dE = 1 \tag{8.50}$$

so that the integrated density is given by

$$n_K^{(Q)}(E;\Delta) = \int_{-2-2\Delta}^{E} D_K^{(Q)}(E';\Delta)\,dE' \quad . \tag{8.51}$$

It is also clear that the present DOS works irrespective of $P^2 \leq (2b+2)^2$ and $\Delta > 0$, so that the extra P-sensitivity displayed in (35) in Hasegawa et al (1990) is superfluous. Working on Riemannian manifolds, there are interesting studies concerning the influence of the surface curvature on DOS-evaluations (Comtet (1987)).

Making the identification $g(E) = D_K^{(Q)}(E;\Delta)$ we are in a position to perform evaluations of Lyapunov exponents with the help of (6.36). Total bandwidth calculations can also be addressed (Thouless (1983a), Wannier (1978)). So far, we have to say that the total bandwidth , i.e. the Lesbeque measure $\mathbb{S}$ of the DOS-support is (Aubry and André (1980), Avron et al (1990))

$$\mathbb{S} = 4(1-\Delta) \tag{8.52}$$

if $\Delta < 1$. In addition, proofs have also been given that the total bandwidth $\mathbb{S}$ behaves as (Thouless (1983b), Tan (1995), Last and Wilkinson (1992))

$$Q\mathbb{S} \rightarrow \frac{32C}{\pi} \tag{8.53}$$

if $Q \rightarrow \infty$ and $\Delta = 1$, where $C \cong 0.915965$ is Catalan's constant . Further applications like thermodynamic properties (Alexandrov and Bratkovsky (2001), Kishigi and Hasegawa (2002)) and the thermodynamic Hall conductance formula (Streda (1982)) are referred shortly in section 8.6.

8.5 Numerical DOS-studies

Doing numerical applications one looks for localized pagoda-like structures both in the energy-and Δ-dependence of the DOS-formulae (Wannier et al

(1979), Hasegawa et al (1990), Papp et al (2002a)). Such structures have to be understood as signatures of the $2DEG$ on a square lattice threaded by a transversal and homogeneous magnetic field. Choosing at the beginning $Q = 3$ and $E = 0$, one finds that the Δ-dependence of $D_K^{(3)}(0;\Delta)$ is characterized by a sole singularity located at $\Delta = 1$, as shown in figure 8.2. This behavior is preserved for arbitrary but odd Q-values, whereas $D_K^{(3)}(0;\Delta) = 0$ if Q is even. If $E \neq 0$ one gets faced with localized peaks, the number of which increases with Q. This is illustrated for $E = 3$ and $Q = 3$ in figure 8.3.

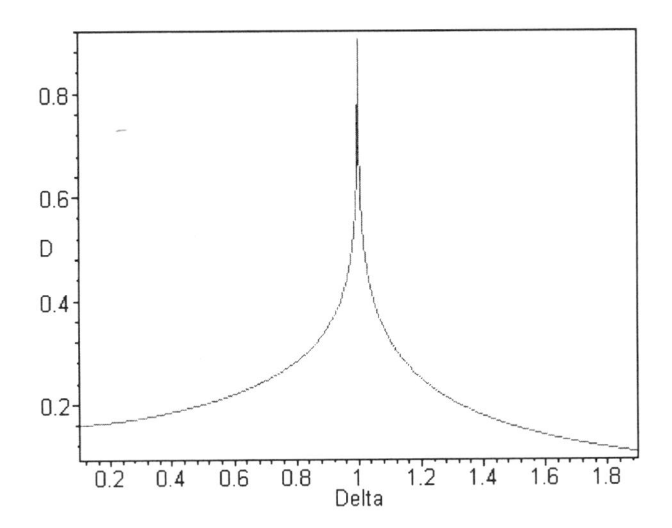

Fig. 8.2 **The Δ-dependence of $D_K^{(3)}(E;\Delta)$ for $E = 0$.**

The influence of the multiplicity parameter $N_s(5) = 2$ is illustrated in figure 8.4, in which one displays the energy dependence of $D_K^{(5)}(E;1)$. Now one has $Q = 5$ and $k = 1$ and 2, which results in a superposition of competing van Hove singularities. It is understood that the complexity of effects just mentioned above increases with Q, as shown e.g. in figure 3 in Hasegawa et al (1990), this time for $Q = 4$ and $\Delta = 0.7$ and 0.1.

8.6 Thermodynamic and transport properties

The energy polynomials as well as the DOS's established above are useful in the study of thermodynamic properties. Using (8.47) and resorting to

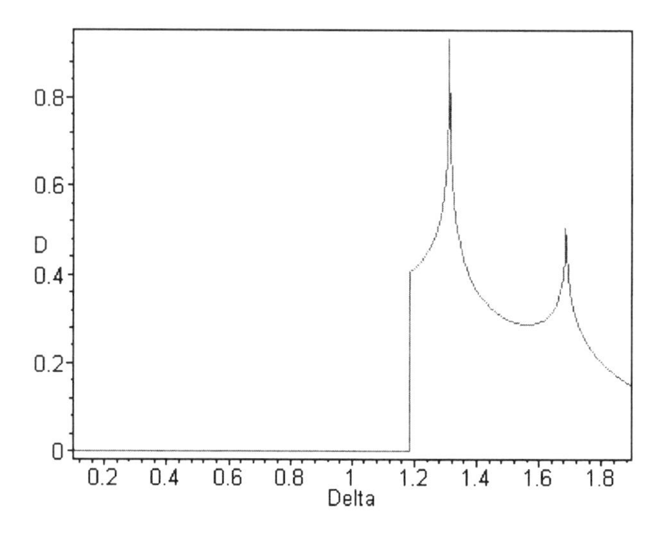

Fig. 8.3 **The Δ-dependence of $D_K^{(3)}(E;\Delta)$ for $E=3$. The two peaks are located at $\Delta \cong 1.31$ and $\Delta \cong 1.68$.**

a fixed Q leads to the thermodynamic potential (see e.g. Alexandrov and Bradkovsky (2001) and Kishigi and Hasegawa (2002)) per unit volume as

$$\Omega(\mu, B) = \Omega_Q^{(\pm)}(\mu, B) = \mp kT \int_{-\infty}^{\infty} dED(E) \ln\left(1 \pm \exp\left(\frac{\mu - E}{kT}\right)\right) \tag{8.54}$$

where μ plays the role of the chemical potential. The "+" ("−") superscript in the front of the r.h.s. of (8.54) corresponds to fermions (bosons). Working in the grand canonical ensemble μ is fixed, in which case the magnetization and the number of particles are given by

$$M_\mu(\mu, B) = -\frac{\partial}{\partial B}\Omega_Q(\mu, B) \tag{8.55}$$

and

$$N_p(\mu, B) = -\frac{\partial}{\partial \mu}\Omega_Q(\mu, B) \tag{8.56}$$

respectively. Then N_p is able to oscillate as a function of B. Fixing, however, the number of particles, one deals with the Helmholtz free-energy.

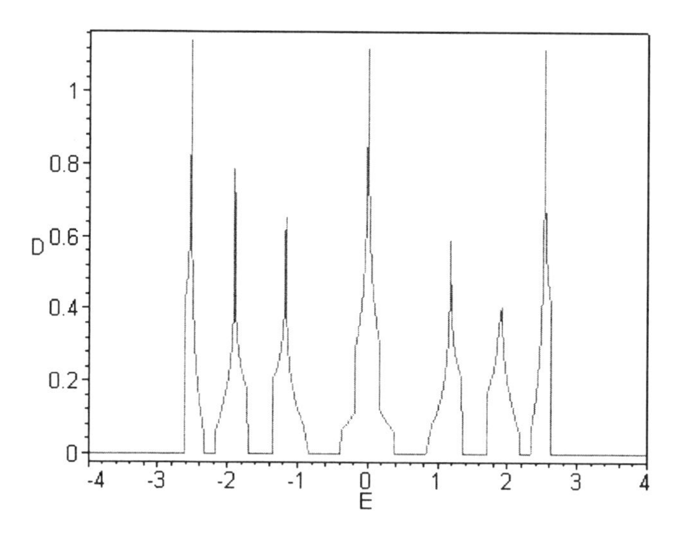

Fig. 8.4 **The energy dependence of $D_K^{(5)}(E;\Delta)$ for $\Delta = 1$.**

$$F_Q\left(N_p, B\right) = \Omega_Q\left(\mu, B\right) + \mu N_p \tag{8.57}$$

in which this time $\mu = \mu\left(N_p, B\right)$, such as given by the canonical ensemble version of (8.56), i.e. by

$$N_p\left(\mu, B\right) = -\frac{\partial}{\partial \mu\left(N_p, B\right)} \Omega_Q\left(\mu\left(N_p, B\right), B\right) \quad . \tag{8.58}$$

The magnetization is then given by

$$M_N\left(N_p, B\right) = -\frac{\partial}{\partial B} F_Q\left(N_p, B\right) \tag{8.59}$$

so that

$$M_N\left(N_p, B\right) = -\left.\frac{\partial}{\partial B} \Omega_Q\left(\mu, B\right)\right|_{\mu=\mu\left(N_p, B\right)} = M_\mu\left(\mu\left(N_p, B\right), B\right) \tag{8.60}$$

by virtue of (8.57).

One sees that the DOS in (8.54) incorporates the single particle states. In general, one starts, however, from the summation formula

$$\Omega(\mu; B) = \mp\frac{1}{\beta}\sum_j \ln(1 \pm \exp(\beta(\mu - E_j))) \tag{8.61}$$

where $\beta = 1/kT$. This in turn can be handled with the help of the Mellin-transform

$$F(t) = \int_0^\infty f(x)x^{t-1}dx \quad . \tag{8.62}$$

Inverting (8.62) yields the original function as

$$f(x) = \frac{1}{2\pi i}\int_{\sigma-i\infty}^{\sigma+i\infty} F(t)\frac{dt}{x^t} \tag{8.63}$$

provided that

$$f(x) \in \{L_2(0,\infty), dx/x^{2\sigma+1}\} \tag{8.64}$$

for $\sigma > \sigma_0$. Then (8.61) can be rewritten as (Grandy and Goulart Rosa (1981))

$$\Omega(\mu; B) = \mp\frac{1}{2i\beta}\sum_j \int_{c-i\infty}^{c+i\infty} \exp(\beta t(\mu - E_j))\frac{\cos^u(\pi t)}{\sin(\pi t)}\frac{dt}{t} \tag{8.65}$$

such that $0 < c < 1$, where $u = 1$ ($u = 0$) for bosons (fermions). This amounts to establish the single particle partition function

$$Z_1(\beta) = \sum_j \exp(-\beta E_j) \tag{8.66}$$

which, in general, is sensitive to the degeneration of energy levels. Thermodynamic signatures of discrete systems can then be easily established combining by resorting to (5.39) and (5.40). This yields e.g. the partition function

$$Z_1(\beta) = \sqrt{q_0}[[N_0 + 1]]_{q_0} \tag{8.67}$$

in accord with (1.15), where

$$q_0 = \exp(-\hbar\beta\omega_0) > 0 \quad , \tag{8.68}$$

has the meaning of a thermodynamic counterpart of the deformation parameter characterizing the harmonic oscillator on the relativistic configuration space (see (11.3)).

It should be noted that DOS-calculations can also be performed with the help of Green-functions (see also Datta (1995)). Indeed, we can define the Green-function relying on a Hamiltonian $\mathcal{H}$ as

$$G_{(\pm)}(E) = \frac{1}{E - \mathcal{H} \pm i\varepsilon} \tag{8.69}$$

in which case

$$\delta(E - \mathcal{H}) = -\frac{1}{2\pi i}\left(G_{(+)}(E) - G_{(-)}(E)\right) \quad . \tag{8.70}$$

The integrated DOS is then given by

$$n(E) = \int_{-\infty}^{E} Tr\delta(E' - \mathcal{H})\, dE' \tag{8.71}$$

which has been also applied in the derivation of the Hall conductance .

Transport properties are also of an actual interest. Accounting for the velocity operator

$$v_j = -\frac{i}{\hbar}[x_i, \mathcal{H}] = \frac{i}{\hbar}\left[x_j, G^{-1}_{(\pm)}\right] \tag{8.72}$$

one obtains the Hall-conductivity in terms of the thermodynamic formula (Streda (1982))

$$\sigma_{xy}(E = E_F, T = 0) = \frac{ec}{S_\perp}\frac{\partial}{\partial B} n(E_F) \tag{8.73}$$

when the Fermi-energy E_F is in a gap, which competes with (7.54). Here

$$n(E_F) = \int_{-\infty}^{E_F} D(E)\, dE \tag{8.74}$$

denotes the number of states below E_F, which provides the total number $n_{tot} = n_s S_\perp$ of electrons within the area $S_\perp$ of the $2D$ sample. At this point let us assume that each magnetic flux quantum is occupied by an integer number, say n_H, of electrons. Accordingly a typical Hall resistivity contribution like $\rho_{xy} = B/ecn_s$ exhibits the quantized form $\rho_{xy} = h/e^2 n_H$, which proceeds in accord with (7.56). These quantized values are seen as plateaus in the B-dependence of ρ_{xy}. It is understood that such relationships are valid at low temperatures and strong magnetic fields. When there are several carriers characterized by the charges e_j (now $e_j > 0$ or $e_j < 0$) with the number densities $n_s^{(j)}$, the absolute value of the Hall-resistivity contribution becomes

$$\rho_{xy} = \frac{\Phi}{cS\perp} \left| \sum_j n_s^{(j)} e_j \right|^{-1} \tag{8.75}$$

where $\Phi = BS_\perp$, while the total number of carriers is

$$n_{tot} = \frac{S\perp}{e} \left| \sum_j n_s^{(j)} e_j \right| = n_H \frac{\Phi}{\Phi_0} \quad . \tag{8.76}$$

The temperature dependence of the Hall conductivity is given by

$$\sigma_{xy}(\mu, T) = \int_{-\infty}^{+\infty} \left(-\frac{\partial}{\partial E} f_D(E) \right) \sigma_{xy}(E, 0)\, dE \tag{8.77}$$

where

$$f_D(E) = \frac{1}{\exp\left((E-\mu)/kT\right) + 1} \tag{8.78}$$

is the Fermi-Dirac distribution . The equivalence between (7.54) and (8.73) proceeds in terms of the Diophantine equation (Kohmoto (1989), Hatsugai (1993))

$$j = Qs_j + Pt_j \tag{8.79}$$

via $n(E) \to j/Q$, where s_j and t_j are integers, whereas $j = 1, 2, ..., Q$, as before. The influence of the spin-orbit interaction on the magnetoconductivity of a $2D$ electron gas has also been analyzed (Wang and Vasilopoulos (2003)).

8.7 The 1D ring threaded by a time dependent magnetic flux

Besides the Harper-equation discussed above, the description of other discretized systems under the influence of the magnetic field is of further interest, too. A such object of study is the electron on an $1D$ ring-shaped nanowire superlattice which is threaded by a time dependent magnetic flux $\Phi(t)$. So one deals with the quantum-mechanical description of a further nanoscale system, namely of the nanoring. Note that to date such nanorings may be synthetized in molecular electronics. Applying Faraday's law

$$\frac{d\Phi}{dt} = -c \oint \vec{E} \cdot d\vec{l} \tag{8.80}$$

yields the electromotive force

$$E_m = -E_\varphi = \frac{R_0}{2c}\frac{dB}{dt} \tag{8.81}$$

where R_0 denotes the radius of the ring. Equation (8.81) can be rewritten equivalently as

$$E_m = \frac{1}{2\pi R_0}\frac{h}{e}\frac{d\beta}{dt} \tag{8.82}$$

where $\beta(t) = \Phi(t)/\Phi_0$ expresses the time dependent number of flux quanta threading the ring. Cylindrical coordinates are used, which also means that the magnetic field is directed along the Oz -axis. However, the only degree of freedom is the angular coordinate φ. The corresponding vector potential is now given by

$$A_\varphi = \frac{1}{2}BR_0 \tag{8.83}$$

where the subscript φ stands for the angular coordinate. Accordingly, the minimal substitution reads

$$p_\varphi^{(op)} \to p_\varphi^{(op)} + \frac{\hbar}{R_0}\beta \tag{8.84}$$

where

$$p_\varphi^{(op)} = -i\frac{\hbar}{R_0}\frac{\partial}{\partial\varphi} \tag{8.85}$$

is the φ -component of the momentum operator.

Assuming for the moment that the periodic superlattice potential is not applied, results in the time dependent Schrödinger equation (Citrin (2004))

$$i\hbar\frac{\partial}{\partial t}\Psi(\varphi,t)=\frac{\hbar^2}{2m_0R_0^2}\left(-i\frac{\partial}{\partial\varphi}+\beta\right)^2\Psi(\varphi,t) \tag{8.86}$$

in which the wavefunction has to satisfy the periodic boundary condition $\Psi(\varphi+2\pi,t)=\Psi(\varphi,t)$. The above wavefunction can also be rescaled as

$$\Psi(\varphi,t)=\exp(-i\beta\varphi)\,\chi(\varphi,t) \tag{8.87}$$

which produces in turn the modified Schrödinger equation

$$i\hbar\frac{\partial}{\partial t}\chi=-\frac{\hbar^2}{2m_0R_0^2}\frac{\partial^2\chi}{\partial\varphi^2}-\hbar\frac{d\beta}{dt}\varphi\chi \quad . \tag{8.88}$$

This equation will be used later as the continuous limit of the lattice formulation. This time the AB-type boundary condition gets reproduces as

$$\chi(\varphi,t)=\exp(-i2\pi eR_0E_m t/\hbar)\,\chi(\varphi+2\pi,t) \quad . \tag{8.89}$$

which isreminiscent to (2.100). Introducing the magnetic quantum number $m=m_\varphi$ yields the normalized solution of (8.86) as

$$\Psi(\varphi,t)=\frac{1}{\sqrt{2\pi}}\exp i\left[m\varphi-\frac{1}{\hbar}\int_0^t dt'\mathcal{E}(t')\right] \tag{8.90}$$

where

$$\mathcal{E}(t)=\mathcal{E}_m(t)=\frac{\hbar^2}{2m_0R_0^2}(m+\beta(t))^2 \tag{8.91}$$

stands effectively for the pertinent time dependent energy.

Concerning the explicit time dependence of the numbers of flux quanta one obtains a linear expression like

$$\beta(t)=\beta_1(t)=\frac{eR_0E_m}{\hbar}t \tag{8.92}$$

if the electromotive force E_m would be independent of time. However, one deals with the sinusouidal modulation

$$\beta(t) = \beta_s(t) = \frac{\beta_0}{\omega} \sin(\omega t) \tag{8.93}$$

when

$$E_m(t) = \frac{\hbar}{R_0 e} \beta_0 \cos(\omega t) \quad . \tag{8.94}$$

One sees that (8.94) reproduces (8.92) when $\omega \to 0$ if $\beta_0 = eR_0E_m/\hbar$, in which case $E_m = E_m(0)$ and $\beta_1(t) = \beta_0 t$.

Next we have to remark that (2.101), i.e. the energy characterizing free-electrons on the 1D ring threaded by a magnetic flux, reproduces (8.91) in terms of the substitution $\beta \to \beta(t)$. This means that the total persistent current produced by (8.91) exhibits the time dependent form

$$I(\Phi) = I_s(t) = \sum_{k=0}^{\infty} s_k \sin(2k+1)\omega t \tag{8.95}$$

in accord with (2.109), (6.97) and (8.93), where

$$s_k = \frac{4I_F}{\pi} \sum_{l=1}^{\infty} \frac{(-1)^{N_l l}}{l} J_{2k+1}(2\pi l \beta_0/\omega) \quad . \tag{8.96}$$

So we found that under the influence of a time dependent magnetic flux like (8.93), the total persistent current in the ring is characterized by the generation of odd higher harmonics, as indicated by (8.95) and (8.96).

It can be easily verified that the generation of such harmonics is also preserved when one considers further modulations like

$$\beta(t) = \beta_c(t) = \beta_c(0) \cos \omega t \quad . \tag{8.97}$$

Indeed, in this latter case one obtains

$$I(\Phi) = I_s(t) = \sum_{k=0}^{\infty} (-1)^k c_k \sin(2k+1)\omega t \tag{8.98}$$

where now

$$c_k = \frac{4I_F}{\pi} \sum_{l=1}^{\infty} \frac{(-1)^{N_l l}}{l} J_{2k+1}(2\pi l \beta_c(0)) \quad . \tag{8.99}$$

8.8 The tight binding description of electrons on the 1D ring

Using the lattice constant a means that the circumference of the ring encompasses a number of N_s sites like

$$N_s = \frac{2\pi R_0}{a} \quad . \tag{8.100}$$

Correspondingly, the angular coordinate gets discretized as

$$\varphi = n\varphi_0 \tag{8.101}$$

where $n = 1, 2, \ldots, N_s$ and $\varphi_0 = 2\pi/N_s$. Choosing $\beta(t) = \beta_1(t)$, one has

$$\beta_1(t) = \frac{N_s}{2\pi}\omega_B t \tag{8.102}$$

where

$$\omega_B = \frac{eaE_m}{\hbar} \tag{8.103}$$

denotes the Bloch frequency. We have to remark that this frequency has also been used before in connection with (6.126).

Our next point is to consider the energy dispersion law (see e.g. Minot (2004))

$$\varepsilon(k_\varphi) = E_0 - \frac{\Delta}{2}\cos(k_\varphi a) \tag{8.104}$$

instead of (8.1), where Δ has the meaning of a band width and where E_0 is a constant parameter. Combining (8.104) with (8.84) produces the Hamiltonian

$$\mathcal{H}_\varphi = E_0 - \frac{\Delta}{2}\cos\frac{2\pi}{N_s}\left(-i\frac{\partial}{\partial\varphi} + \beta\right) \tag{8.105}$$

which yields the time dependent Schrödinger equation

$$\mathcal{H}_\varphi \Phi(\varphi, t) = i\hbar\frac{\partial}{\partial t}\Phi(\varphi, t) \quad . \tag{8.106}$$

One would then obtain the second order discrete equation (see also Niu (1989))

$$E_0 \Phi_n(t) - \frac{\Delta}{4}\left[e^{i\varphi_0 f}\Phi_{n+1}(t) + e^{-i\varphi_0 f}\Phi_{n-1}(t)\right] = \tag{8.107}$$
$$= i\hbar \frac{\partial}{\partial t}\Phi_n(t)$$

where $\Phi_n(t) = \Phi(n\varphi_0, t)$. Next let us rescale this wavefunction as

$$\Phi_n(t) = \chi_n(t)\exp(-in\varphi_0 f) \quad . \tag{8.108}$$

Then (8.107) becomes

$$(E_0 - n\hbar\omega_B)\chi_n(t) - \frac{\Delta}{4}\left[\chi_{n+1}(t) + \chi_{n-1}(t)\right] = \tag{8.109}$$
$$= i\hbar \frac{\partial}{\partial t}\chi_n(t)$$

which is similar to (6.52), i.e. to the discrete equation characterizing the $1D$ conductor under the influence of a dc electric field. Next we are ready to perform the continuous limit of (8.104), which proceeds in terms of large N_s values. Proceeding to second $1/N_s$ order, one finds that (8.109) reproduces (8.88) if

$$E_0 = \frac{\Delta}{2} = \frac{\hbar^2}{ma^2} \tag{8.110}$$

which is a reasonable result. Moreover, (8.109) is exactly solvable, which is also reminiscent to (6.52). Now one finds (Citrin (2004))

$$\chi_n(t) = \frac{1}{\sqrt{N_s}}\exp\left(-\frac{i}{\hbar}E_0 t\right) \cdot \tag{8.111}$$
$$\sum_{\nu=-\infty}^{+\infty} J_{\nu-n}(z)\exp i\nu\Omega_B(t)$$

where $z = \Delta/2\hbar\omega_B$ and $\Omega_B(t) = \omega_B t + 2\pi m_\varphi/N_s$. It can also be easily verified that $\chi_n(t)$ can be rewritten as

$$\chi_n(t) = \frac{1}{\sqrt{N_s}}\exp i\left[-\frac{E_0 t}{\hbar} + n\Omega_B(t) + z\sin\Omega_B(t)\right] \tag{8.112}$$

which proceeds in accord with (6.97). The AB-type boundary condition needed, namely

$$\chi_{n+N_s}(t) = \exp(2\pi i\beta_1)\,\chi_n(t) \tag{8.113}$$

is also fulfilled, which corresponds to (8.89). It is also clear that

$$\widetilde{\chi}_n(t + N_B T_B) = \widetilde{\chi}_n(t) \equiv \chi_n(t)\exp\left(it\frac{E_0}{\hbar}\right) \tag{8.114}$$

where $T_B = 2\pi/\omega_B$ denotes the Bloch-period and where N_B is an integer. The time periodicity implied in this manner serves as a signature of magnetic Bloch oscillations.

Next let us consider the $\beta = \beta_s(t)$-choice. This amounts to replace ω_B by

$$\omega_B(t) = \omega_B \cos(\omega t) \tag{8.115}$$

in (8.109). So far we would like just to remark that (8.112) is able to satisfy the updated form of (8.109), but only if Δ would be replaced by

$$\Delta(t) = \Delta\cos(\omega t) \tag{8.116}$$

in both equations. Under such conditions the z-parameter remains invariant:

$$z = \frac{1}{2\hbar}\frac{\Delta(t)}{\omega_B(t)} = \frac{1}{2\hbar}\frac{\Delta}{\omega_B} \quad . \tag{8.117}$$

This results in the implementation of a periodic time dependent Hamiltonian, which deserves further attention in terms of the Floquet-approach. In addition, (8.114) is also preserved in so far as ω is an integral multiple of ω_B.

8.9 The persistent current for the electrons on the 1D discretized ring at $T = 0$

Let us consider a discretized 1D ring with the circumference $L = N_s a$, which is pierced by an axial magnetic flux Φ. As in the case of the continuous ring discussed in section 2.3, we shall assume that the electrons move in a field-free space. This leads to the discrete equation

$$-V_0(\psi_{m+1}+\psi_{m-1})=E\psi_m \tag{8.118}$$

working in conjunction with (2.94), i.e. with the AB-type periodic boundary condition

$$\psi_{m+N_s}=e^{2\pi i\beta}\psi_m \quad . \tag{8.119}$$

Using again the wavefunction ansatz

$$\psi_m=\frac{1}{\sqrt{N_s}}\exp{(ikm)} \tag{8.120}$$

gives the dimensionless wave number as

$$k=\frac{2\pi}{N_s}(\beta+n) \tag{8.121}$$

where $n=0,\pm1,\pm2,\dots$, so that the energy one looks for reads

$$E=E_n=-2V_0\cos\frac{2\pi}{N_s}(\beta+n) \quad . \tag{8.122}$$

Applying (2.76) yields the corresponding persistent current as

$$I_n=-\frac{2eV_0}{N_s\hbar}\sin\frac{2\pi}{N_s}(\beta+n) \tag{8.123}$$

which reproduces (2.103) via $N_s\to\infty$, provided that

$$V_0=\frac{\hbar}{2m_0a^2} \quad . \tag{8.124}$$

It is then clear that

$$\frac{2eV_0}{N_s\hbar}=\frac{N_s}{\pi N_e}I_F \tag{8.125}$$

where I_F is done in (2.104).

Now the Fermi energy reads

$$E_F=-2V_0\cos\left(\pi\frac{N_e}{N_s}\right) \tag{8.126}$$

in accord with (2.95). Then the energies less than the Fermi-one have to be handled in terms of the inequality

$$\cos\frac{2\pi}{N_s}(n+\beta) \geqq \cos\left(\pi\frac{N_e}{N_s}\right) \tag{8.127}$$

which expresses the discrete counterpart of (2.106). Making the summations needed gives the total persistent currents at $T=0$ as

$$I(\Phi) = -I_F f_{\pm}(\beta) \tag{8.128}$$

where the "+" and "−" subscripts correspond to even and odd N_e-values, respectively. Accordingly (see also Cheung et al (1988))

$$f_+(\beta) = \frac{N_s}{\pi N_e}\left[\sin\left(\frac{2\pi}{N_s}\beta\right)\sin\left(\frac{N_e\pi}{N_s}\right)\cot\left(\frac{\pi}{N_s}\right)+\right. \tag{8.129}$$

$$\left.+\frac{1}{2}\sin\frac{2\pi}{N_s}\left(\beta-\frac{N_e}{2}\right)+\frac{1}{2}\sin\frac{2\pi}{N_s}\left(\beta-\frac{N_e}{2}-1\right)\right]$$

while

$$f_-(\beta) = \frac{N_s}{\pi N_e}\sin\left(\frac{2\pi}{N_s}\beta\right)\left[1+\cos\frac{\pi}{N_s}(N_e+1)+\right. \tag{8.130}$$

$$\left.+\frac{2}{\sin\left(\frac{\pi}{N_s}\right)}\cos\left(\frac{N_e+1}{2N_s}\pi\right)\sin\left(\frac{N_e-1}{2N_s}\pi\right)\right] \quad .$$

For this purpose the summation formula

$$\sum_{k=1}^{n}\cos(kx) = \cos\left(\frac{n+1}{2}x\right)\sin\left(\frac{nx}{2}\right)\csc\left(\frac{x}{2}\right) \tag{8.131}$$

has been applied (see 1.342.2 in Gradshteyn and Ryzhik (1965)). Performing the $N_s \to \infty$-limit, it can be readily verified that (8.127) and (8.128) reproduce (2.107) and (2.108), respectively. So we found reasonably a discrete generalization of the total persistent current established in section 2.9 for electrons on the usual 1D ring. However, in the present case the current exhibits an enhanced sensitivity with respect to the parameters β, N_e and N_s. The generation of higher harmonics can also be easily established by inserting periodic $\beta(t)$-functions into (8.127) and (8.128).

Chapter 9

The q-Symmetrized Harper Equation

The q-symmetrized Harper-equation (qSHE) deals with the middle band description of Bloch-electrons on the square lattice threaded by a transversal and homogeneous magnetic field. This equation has received a special interest in connection with novel mathematical developments concerning q-deformed symmetries and second order q-difference equations (Wiegmann and Zabrodin (1994a), Wiegmann and Zabrodin (1994b), Hatsugai et al (1994), Faddeev and Kashaev (1995), Abanov et al (1998), Krasovsky (1999)). Explicit solutions have also been discussed in some more detail for fixed Q-values (Papp and Micu (2002b)), but non-polynomial generalizations towards arbitrary values of the anisotropy parameter Δ have also been proposed (Papp (2003)). Several aspects of this interesting matter are presented below.

9.1 The derivation of the generalized qSHE

Let us consider the Coulomb-gauge expressed by the vector-potential

$$A_l = (-1)^l \frac{B}{2}(x_1 + x_2 + \alpha_l a) \equiv (-1)^l \frac{B}{2}\mathcal{X}_l \tag{9.1}$$

instead of the Landau-gauge used before in terms of (8.2), where α_l ($l = 1, 2$) are, for the moment, arbitrary gauge parameters. In addition, we shall work by using the wavefunction ansatz

$$\psi(\overrightarrow{x}) = \exp\left(i\overrightarrow{k}\cdot\overrightarrow{x}\right)\varphi(\overrightarrow{x}) \tag{9.2}$$

where $\overrightarrow{k}\cdot\overrightarrow{x} = k_1x_1 + k_2x_2$ and

$$\varphi_B\left(\overrightarrow{x}\right) = \varphi\left(x_1 + x_2\right) \tag{9.3}$$

which differs, of course, from (8.6). One recognizes that (9.3) serves to the very implementation of the $1D$ description. The minimal substitution, such as done by (8.3) will also be applied, but now in terms of (9.1). For this purpose, we can start from the energy-dispersion law expressed by (8.1), but reversing the roles of θ_1 and θ_2, the alternative formula

$$\mathcal{E}_d^{(1)}\left(\overrightarrow{\theta}\right) = \Delta\cos\theta_1 + \cos\theta_2 \tag{9.4}$$

can be invoked as well. Furthermore, we have to apply the Baker-Campbell-Hausdorff formula (see e.g. (3.1.20) in Louisell (1973))

$$\exp\left(A\right)\exp\left(B\right) = \exp\left(A+B\right)\exp\left(\frac{1}{2}\left[A,B\right]\right) \tag{9.5}$$

which gives

$$\exp\left(a\frac{\partial}{\partial x_l} + \frac{i}{2}\left(-1\right)^l \hbar^* \mathcal{X}_l\right) = \tag{9.6}$$

$$\exp\left(\frac{i}{4}\left(-1\right)^l \hbar^*\right)\exp\left(\frac{i}{2}\left(-1\right)^l \mathcal{X}_l\right)\exp\left(a\frac{\partial}{\partial x_l}\right)$$

for $l = 1, 2$, where $\hbar^* = 2\pi\beta$, as indicated by (7.5). Finally, we have to perform translations such as done by (9.6), as well as discretization $x_l = n_l a$. Then $x_1 + x_2 = na$ and $n = n_1 + n_2$, where n_1 and n_2 are arbitrary integers. Putting together these results yields the second order discrete equation

$$\left(\Delta\frac{\exp\left(i\theta_1\right)}{q^{n+\alpha_1+1/2}} + \frac{q^{n+\alpha_2+1/2}}{\exp\left(-i\theta_2\right)}\right)\varphi_{n+1} + \left(\Delta\frac{q^{n+\alpha_1-1/2}}{\exp\left(i\theta_1\right)} + \frac{\exp\left(-i\theta_2\right)}{q^{n+\alpha_2-1/2}}\right)\varphi_{n-1} = E\varphi_n \tag{9.7}$$

by virtue of (9.4), where $\varphi_n = \varphi_B\left(na\right)$ and where one has

$$q = \exp\left(\frac{i}{2}\hbar^*\right) \tag{9.8}$$

in accord with (7.13). Using the wavefunction

$$\psi(z) = \sum_{n=-\infty}^{+\infty} \varphi_n z^n \tag{9.9}$$

it can be easily verified that (9.7) leads to the q-difference equation

$$\left(\frac{q^{\alpha_2-1/2}}{z\exp(-i\theta_2)} + z\frac{\Delta q^{\alpha_1+1/2}}{\exp(i\theta_1)} \right) \psi(qz) + \tag{9.10}$$

$$+ \left(z\frac{\exp(-i\theta_2)}{q^{\alpha_2+1/2}} + \frac{\Delta\exp(i\theta_1)}{zq^{\alpha_1-1/2}} \right) \psi\left(\frac{z}{q}\right) = E\psi(z) \quad .$$

After having been arrived at this stage, let us insert $\alpha_1 = \alpha_2 = 1/2$ and $\theta_2 = -\theta_1 = \pi/2$ into (9.7). This yields

$$i\left(\frac{1}{z} + \Delta qz\right)\psi(qz) - i\left(\frac{z}{q} + \frac{\Delta}{z}\right)\psi\left(\frac{z}{q}\right) = E(\Delta)\psi(z) \tag{9.11}$$

which reproduces the well known qSHE (Wiegmann and Zabrodin (1994a))

$$\mathcal{H}_q\psi(z) \equiv i\left(\frac{1}{z} + qz\right)\psi(qz) - i\left(\frac{z}{q} + \frac{1}{z}\right)\psi\left(\frac{z}{q}\right) = E(1)\psi(z) \tag{9.12}$$

as soon as $\Delta = 1$. Starting however from (8.1) and repeating the same steps gives the complementary equation

$$i\left(\frac{\Delta}{z} + qz\right)\psi(qz) - i\left(\frac{1}{z} + \Delta\frac{z}{q}\right)\psi\left(\frac{z}{q}\right) = E_1(\Delta)\psi(z) \tag{9.13}$$

instead of (9.11), such that

$$E_1(\Delta) = \Delta E\left(\frac{1}{\Delta}\right) \quad . \tag{9.14}$$

Under such circumstances we have to realize that both (9.11) and (9.13) can be viewed as $\Delta \neq 1$ -generalizations of the qSHE. For convenience, we can restrict, however, to (9.11) only, such as considered before (Papp (2003)). So far, mutual conversions of (9.12) and (8.7) have been done for $\Delta = 1$ only (see Appendix A in Krasovsky (1999)). Note, however, that (8.9) exhibits the q-difference form

$$\left(\frac{\exp(i\theta_1)}{z} + z\exp(-i\theta_1) \right) u(z) + \tag{9.15}$$

$$+\Delta \exp(i\theta_2)\, u\left(q^2 z\right) + \Delta \exp(-i\theta_2)\, u\left(\frac{z}{q^2}\right) = Eu(z)$$

as well as the dual partner

$$\exp(i\theta_1)\, f\left(\frac{z}{q^2}\right) + \exp(-i\theta_2)\, f\left(q^2 z\right) + \frac{\Delta}{z}\exp(i\theta_2)\, f(z) + \qquad (9.16)$$

$$\frac{\Delta z}{\exp(i\theta_2)} f(z) = Ef(z)$$

for which (7.57) has been used. The wavefunction quotations are selfconsistently understood in terms of (9.9).

However, there is still a point which has to be clarified. Indeed, inserting $\theta_1 = \theta_{10} = -\pi/2$ and $\theta_2 = \pi/2$ into the r.h.s. of (7.39) yields

$$f\left(-\frac{\pi}{2}, \frac{\pi}{2}\right) = 0 \qquad (9.17)$$

if $t_c = 0$, but for odd Q-values only. This would then ensure the middle band description referred to above. Proofs have also been given that both virial and Hellmann-Feynman theorems can be applied reasonably to (9.11) (Micu and Papp (2003)).

9.2 The three term recurrence relation

Rescaling the energy

$$E = i\left(q - \frac{1}{q}\right) W \qquad (9.18)$$

it can be easily verified that (9.12) can be rewritten equivalently as

$$\left(\mathcal{D}_z^{(q)} + z\mathcal{D}_z^{(q)} z\right)\psi(z) = W\psi(z) \qquad (9.19)$$

where $\mathcal{D}_z^{(q)} = D_q/D_q z$, by virtue of (1.16). This in turn can be solved in terms of three term recurrence relations. Indeed, inserting the polynomial wavefunction

$$\psi(z) = \psi_q^{(Q)}(z) = \sum_{n=0}^{Q-1} C_n z^n \qquad (9.20)$$

into (9.19) yields the recurrence relation

$$[n+1]_q\, C_{n+1} + [n]_q\, C_{n-1} = W C_n \tag{9.21}$$

where $C_0 = 1$. On the other hand one has $[Q]_q = 0$ by virtue of (7.4) and (9.8), in which case

$$C_{Q+1} = C_{Q+2} = \ldots = 0 \quad . \tag{9.22}$$

This shows that the energy levels corresponding to a fixed value of the Q-parameter should be established via

$$C_Q = C_Q\,(q, W) = 0 \quad . \tag{9.23}$$

On the other hand there is

$$C_Q\,(q, W) = q^{N_Q} \frac{f_Q\,(q^2, W)}{[[Q]]_{q^2}!} \tag{9.24}$$

where $f_Q\,(q^2, W)$ denotes a polynomial of degree Q in W and where

$$N_Q = \frac{Q}{2}\,(Q-1) \quad . \tag{9.25}$$

Equation (9.24) comes from reasonable generalizations of some few explicit results. We have to realize, within the same context, that $f_Q(q^2, W) = f_Q(1/q^2, W)$, which means in turn that $C_Q(q, W) = C_Q(1/q, W)$and $W(q^2) = W(1/q^2)$. Conversely, this latter equality implies the former one by virtue of (9.21). In other words the wavefunction itself is invariant under $q \to 1/q$, i.e. $\psi_q^{(Q)}(z) = \psi_{1/q}^{(Q)}(z)$. One would then have the mappings $E \equiv E_q \to E_{1/q} = -E_q$, which proceeds in accord with (9.21) and (9.18).

Now what remains is to insert (9.24) into (9.21). This yields the symmetrized recurrence relations

$$f_Q\,(q^2, W) = W f_{Q-1}\,(q^2, W) - \Omega_{Q-2}^2 f_{Q-2}\,(q^2, W) \tag{9.26}$$

for $Q = 1, 2, 3, \ldots$, where

$$\Omega_{Q-2} = q^{2-Q}\,[[Q-1]]_{q^2} \quad . \tag{9.27}$$

Using the combination

$$\Gamma_n = \Gamma_n(q) = \Gamma_n(1/q) = q^n + \frac{1}{q^n} = 2\cos\left(n\frac{\hbar^*}{2}\right) \tag{9.28}$$

which is invariant under $q \to 1/q$, it can be easily proved that

$$\Omega_{Q-2}(q) = \Omega_{Q-2}(1/q) = \begin{cases} \Gamma_1 + \Gamma_3 + \ldots + \Gamma_{Q-2}\,, & Q = \text{odd} \\ 1 + \Gamma_2 + \ldots + \Gamma_{Q-2}\,, & Q = \text{even}\,. \end{cases} \tag{9.29}$$

Under such circumstances one obtains the eigenvalue equation

$$f_Q\left(q^2, W\right) = f_Q\left(1/q^2, W\right) = 0 \tag{9.30}$$

by virtue of (9.22) and (9.24), which produces precisely a number of Q real W-roots, say

$$W = W_j^{(Q)}\left(q^2\right) = W_j^{(Q)}\left(1/q^2\right) \tag{9.31}$$

where $j = 1, 2, \ldots, Q$. The q-normalization of present wavefunction can also be done in terms of (1.32). For this purpose we can choose $z \in [-1, 1]$, but other normalization-intervals like $z \in [0, b]$ can also be invoked.

The first six f_Q polynomials are given by (Papp and Micu (2002b))

$$f_1\left(q^2, W\right) = W \tag{9.32}$$

$$f_2\left(q^2, W\right) = W^2 - 1 \tag{9.33}$$

$$f_3\left(q^2, W\right) = W\left(W^2 - 3 - \Gamma_2\right) \tag{9.34}$$

$$f_4\left(q^2, W\right) = W^4 - 6W^2 + 3 + \tag{9.35}$$

$$+ \left(2 - 3W^2\right)\Gamma_2 + \left(1 - W^2\right)\Gamma_4$$

$$f_5\left(q^2, W\right) = W\left[W^4 - 10W^2 + 21 + \right. \tag{9.36}$$

$$+ \left(17 - 6W^2\right)\Gamma_2 + \left(11 - 3W^2\right)\Gamma_4 +$$

$$+\left(5-W^2\right)\Gamma_6+\Gamma_8\Big]$$

and

$$f_6\left(q^2,W\right)=W^6-15W^4+81W^2-37+ \tag{9.37}$$
$$+\left(71W^2-10W^4-34\right)\Gamma_2++\left(53W^2-6W^4-27\right)\Gamma_4+$$
$$+\left(33W^2-3W^4-18\right)\Gamma_6++\left(16W^2-W^4-10\right)\Gamma_8+$$
$$+\left(5W^2-4\right)\Gamma_{10}+\left(W^2-1\right)\Gamma_{12}\quad .$$

The above energies reproduce the ones obtained before for $Q=1-6$ (see (8.24)-(8.287)) in terms of (9.18) as soon as $P=P_s$-discrete realizations are accounted for. Further cases like $Q>7$ remain to be solved numerically. It should be stressed, however, that explicit energy results established in this way are useful in order to probe several conjectures concerning the spectrum of the qSHE.

9.3 Symmetry properties

We have to realize that f_Q can be represented as

$$f_Q\left(q^2,W\right)=W^{\alpha_Q}\sum_{n=0}^{\beta_Q}d_n^{(Q)}\left(W^2\right)\Gamma_{2n} \tag{9.38}$$

in which $d_n^{(Q)}\left(W^2\right)$ are constituent polynomials in W^2. Furthermore

$$\alpha_Q=0\ ,\qquad \beta_Q=\frac{1}{4}Q\left(Q-2\right) \tag{9.39}$$

for even Q values, whereas

$$\alpha_Q=1\ ,\qquad \beta_Q=\frac{1}{4}\left(Q-1\right)^2 \tag{9.40}$$

for odd Q values. So one finds

$$d_{\beta_Q}^{(Q)}\left(W^2\right)=\left(-1\right)^{Q/2}\left(1-W^2\right) \tag{9.41}$$

and

$$d_{\beta_Q}^{(Q)}\left(W^2\right) = (-1)^{(Q-1)/2} \tag{9.42}$$

respectively. It is also clear that $d_0^{(Q)}\left(W^2\right)$ is a polynomial of degree $(Q - \alpha_Q/2)$ in W^2, but the general description of remaining constituents is still an open problem.

Accounting for (9.18) and (9.31) yields the energy spectrum

$$S_Q[\hbar^*] = \{E_j^{(Q)}(\hbar^*) \mid \quad j = 1, 2, ..., Q\} \tag{9.43}$$

for $Q = 1, 2, 3, \ldots$, where

$$E = E_j^{(Q)}(\hbar^*) = -2\sin\frac{\hbar^*}{2} W_j^{(Q)}(q^2) \tag{9.44}$$

and where we shall assume hereafter that $\hbar^* \in [0, 2\pi]$. This continuous extrapolation is able to serve for a better description of underlying symmetries. For the sake of discrimination we shall then put $\widetilde{x} = \hbar^*$, thereby considering $\widetilde{x}$ as a continuous variable. The actual discrete spectrum of the qSHE is then given by virtue of the intersection

$$\mathcal{E}_Q = S_Q\left[\widetilde{x}\right] \bigcap M_Q \quad . \tag{9.45}$$

which shows that we have to resort to the set of crossing points between the $\widetilde{x}$ -dependent energy curves belonging to $S_Q\left[\widetilde{x}\right]$ and and the set M_Q consisting of vertical lines like $\widetilde{x} = \widetilde{x}_s$, where $\widetilde{x}_s = 2\pi P_s/Q$.

Further inter-connections with the usual Harper-equation (8.7) are also worthy of being mentioned. Indeed, inserting $P = P_s \equiv P_k^{(Q)}$ into (9.30) yields a number of $N_s(Q)$ discrete polynomial realizations like $P_Q^{(k)}(E) = 0$, where $P_Q^{(k)}(E) \equiv P_Q^{(k)}(E;1)$ is a polynomial of degree Q in E, as shown before. Such realizations have to be established in terms of a subsequent normalization, which proceeds by choosing the coefficient of E^Q to be unity. Accordingly

$$P_Q^{(k)}(E) = \left(i\left(q - \frac{1}{q}\right)\right)^Q f_Q\left(q^2, W\right)\Big|_{q=\exp\left(i\pi P_k^{(Q)}/Q\right)} \tag{9.46}$$

which works in combination with (9.18). These polynomials are precisely the ones produced by applying either the secular equation method or, equivalently, the transfer matrix technique to (8.7).

The present energies are well ordered in the sense that following inequalities

$$E_1^{(Q)}(\widetilde{x}) \leq E_2^{(Q)}(\widetilde{x}) \leq \ldots \leq E_j^{(Q)}(\widetilde{x}) \tag{9.47}$$

are valid irrespective of $\widetilde{x} \in [0, 2\pi]$. This is synonymous with a non crossing behavior, which means that one has just contact points $\widetilde{x} = \widetilde{x}_C^{(Q)}(E)$ corresponding to the equality signs in (9.47). One realizes that the $\widetilde{x}$-derivatives of energy eigenvalues are not continuous in such contact points. Next there is

$$E_j^{(Q)}(\widetilde{x}) = -E_{Q-j+1}^{(Q)}(\widetilde{x}) \tag{9.48}$$

which exhibits the so called energy reflection symmetry (Shifmann and Turbiner (1999)). This proceeds in accord with the underlying $SL_q(2)$ - symmetry. In addition

$$E_j^{(Q)}(0) = E_j^{(Q)}(2\pi) = 0 \tag{9.49}$$

but, excepting the zero energy solution, there is

$$E_j^{(Q)}(\pi) = \pm 2 \quad . \tag{9.50}$$

Correspondingly, $\widetilde{x} = \pi$ stands for a symmetry axis of the spectrum

$$E_j^{(Q)}(\widetilde{x}) = E_j^{(Q)}(2\pi - \widetilde{x}) \tag{9.51}$$

which holds for $0 < \widetilde{x} < 2\pi$. This also means that $2\pi - \widetilde{x}_C^{(Q)}(E)$ is a contact point as soon as $\widetilde{x}_C^{(Q)}(E)$ does it. If $Q = 4$, the contact points are $\widetilde{x}_{c,1}^{(4)}(0) = 2\pi/3$, $\widetilde{x}_{c,2}^{(4)}(0) = 4\pi/3$, and

$$\widetilde{x}_{c,1}^{(4)}(\pm 2) = \pi \tag{9.52}$$

as shown in figure 9.1.

The $\widetilde{x}$-dependence of the five energy levels $E = E_j^{(5)}(\widetilde{x})$ corresponding to $Q = 5$ is displayed in figure 9.2. The $Q = 7$ patterns are actually even more sophisticated, as shown in figure 9.3.

Having obtained $f_Q\left(q^2, W\right)$ opens the way to establish B-derivatives of energy eigenvalues via

$$\frac{2\pi\hbar}{ea^2}\frac{\partial}{\partial B} = i\frac{\pi}{Q}q\frac{\partial}{\partial q} \tag{9.53}$$

where q stands, of course, for $\exp(i\widetilde{x}/2)$ and where Q is fixed. This latter equation is useful in the study of the magnetization (see (8.55) and (8.59)) as well as of the Hall conductance (see (8.73)).

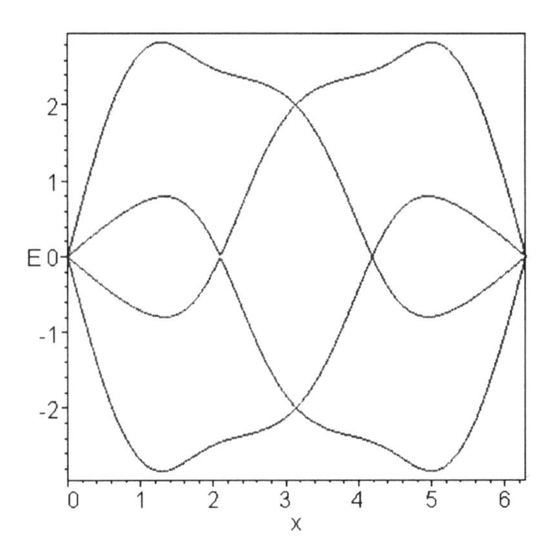

Fig. 9.1 **The $x = \tilde{x}$-dependence of the four energy levels $E = E_j^{(4)}(\tilde{x})$, where $j = 1, 2, 3$ and 4. One sees that $N_C^{(4)}(0) = 2$, while $N_C^{(4)}(2) = N_C^{(4)}(-2) = 1$.**

9.4 The $SL_q(2)$-symmetry of the q SHE

Let us consider two operators B and C for which

$$B\psi(z) = \frac{z}{q - q^{-1}} \left(q^{2j} \psi\left(\frac{z}{q}\right) - q^{-2j} \psi(qz) \right) \tag{9.54}$$

and

$$C\psi(z) = \mathcal{D}_z^{(q)} \psi(z) = \frac{\psi(qz) - \psi(q^{-1}z)}{z(q - q^{-1})} \tag{9.55}$$

which proceeds in accord with (1.16). It is understood j plays the role of a quantum number of the angular momentum, or equivalently, of the spin. This relies on a representation having the dimension $2j + 1$.

The Hamiltonian characterizing the qSHE such as done by (9.12) can then be expressed as

$$\mathcal{H}_q = i(q - q^{-1})(C - q^Q B) \tag{9.56}$$

as soon as

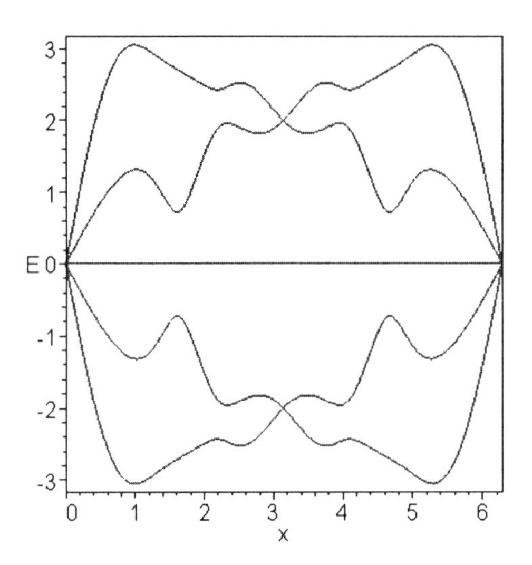

Fig. 9.2 **The $x = \widetilde{x}$-dependence of the five energy levels$E = E_j^{(5)}(\widetilde{x})$, where $j = 1, 2, ..., 5$. The only $E = 0$ contact points are located at $\widetilde{x} = 0$ and $\widetilde{x} = 2\pi$.**

$$q^{2j+1} = q^Q \quad . \tag{9.57}$$

This shows that the j-spin can be established as

$$j = \frac{Q-1}{2} \tag{9.58}$$

with the understanding that the general solution is $j = (Q-1)/2$ (mod Q). The related representation concerns the space of $\psi(z)$-polynomials of degree $2j$, which agrees with (9.20). We then have to realize that a such symmetry (Wiegmann and Zabrodin (1994a)) is produced by the quantum group $SL_q(2)$. More exactly, one resorts to two additional generators, say A and D, such that

$$A\psi(z) = q^{-j}\psi(qz) \tag{9.59}$$

and

$$D\psi(z) \equiv A^{-1}\psi(z) = q^j\psi\left(\frac{z}{q}\right) \quad . \tag{9.60}$$

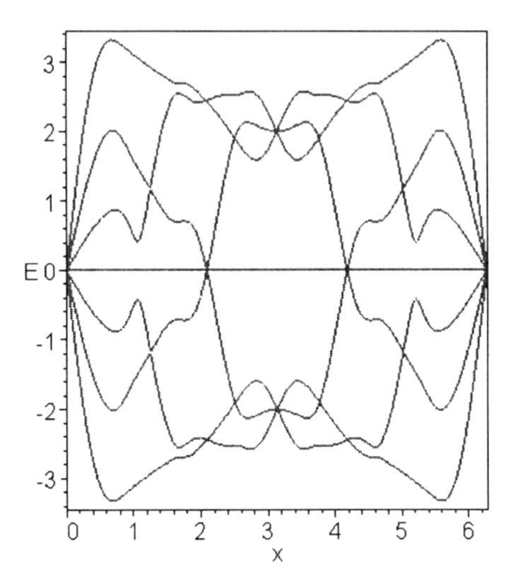

Fig. 9.3 **The $x = \widetilde{x}$ -dependence of the seven energy levels $E = E_j^{(7)}(\widetilde{x})$, where $j = 1, 2, ..., 7$. One has $N_C^{(7)}(0) = 2$ and $N_C^{(7)}(2) = N_C^{(7)}(-2) = 1$, but $N_C^{(7)}(E_i) = N_C^{(7)}(-E_i) = 2$ for $E_1 \cong 1.360$, $E_2 \cong 1.902$ and $E_3 \cong 2.119$.**

Putting together the above generators yields the relationships characterizing the $SL_q(2)$-group just referred to above, namely (Faddeev et al (1990), Wiegmann and Zabrodin (1994a))

$$AB = qBA \ , \qquad BD = qDB \tag{9.61}$$

$$DC = qCD \ , \qquad CA = qAC \tag{9.62}$$

and

$$[B, C] = \frac{A^2 - D^2}{q - q^{-1}} \tag{9.63}$$

in which $AD = 1$. The Casimir-operator of this group is given by

$$\Omega_q = \frac{q^{-1}A^2 + qD^2}{(q - q^{-1})^2} + BC - \frac{2}{(q - q^{-1})^2} \tag{9.64}$$

so that

$$\Omega_q z^n \equiv \omega_q z^n = [j + \frac{1}{2}]_q^2 z^n \quad . \tag{9.65}$$

Accounting for (9.58), one realizes that the Casimir-eigenvalue ω_q vanishes if P is even, but

$$\omega_q = -\frac{4}{(q - q^{-1})^2} \tag{9.66}$$

if P is odd. The Q-denominator can then be either even or odd.

Furthermore one has

$$A^2 = \exp\left(i\hbar^*\left(-j + z\frac{d}{dz}\right)\right) \tag{9.67}$$

so that

$$\frac{A^2 - D^2}{q - q^{-1}} \to 2S_3 \equiv 2\left(-j + z\frac{d}{dz}\right) \tag{9.68}$$

if $q \to 1$. Similarly,

$$C \to S_- = \frac{d}{dz} \tag{9.69}$$

and

$$B \to S_+ = z\left(2j - z\frac{d}{dz}\right) \quad . \tag{9.70}$$

One would then obtain the relationships

$$[S_3, S_\pm] = \pm S_\pm \tag{9.71}$$

and

$$[S_+, S_-] = 2S_3 \tag{9.72}$$

which express the generator-algebra of the classical group $SL(2)$. Note that (9.72) reproduces (5.24) up to the sign, whereas (5.23) and (9.71) are equivalent. Accordingly

$$\Omega_q \to S_+S_- + S_3(S_3 - 1) \tag{9.73}$$

whereas $\omega_q \to j\,(j+1) + 1/4$. It has been shown before (see section 5.2) that the group $SL\,(2)$ is useful in order to discuss partial algebraizations of spectral problems (Shifman and Turbiner (1999), Shifman and Turbiner (1989b)), but solutions to novel periodic Hamiltonians can also be established (Bagchi and Ganguly (2003)). We have also to remark that the above generators reproduce the $SL(2)$ commutation relations done by (5.23) and (5.24) via

$$J_n^- = S_- \tag{9.74}$$

$$J_n^0 = S_3 \tag{9.75}$$

and

$$J_n^+ = -S_+ \tag{9.76}$$

where $j = n/2$. Of course, this results in differential realizations acting in the space of $P_n\,(z)$-polynomials of degree at most n in the variable z.

9.5 Magnetic translations

Speaking about magnetic translations means that the Hamiltonian of the Harper-equation, or of the q-symmetrized one, is expressed by a sum of four generators, say T_i $(i = 1, 2, 3, 4)$ exhibiting relationships like (Fradkin (1991), Wiegmann and Zabrodin (1994a), Zak (1964), Boon (1972)),

$$T_1 T_2 = q^2 T_2 T_1 \tag{9.77}$$

and

$$T_1 T_4 = \frac{1}{q^2} T_4 T_1 \quad . \tag{9.78}$$

Such relationships are typical for so-called ray representations of magnetic translations. Choosing e.g. (9.11), one would then have the realizations

$$T_1 = \frac{i}{z} q^{J_3} \tag{9.79}$$

$$T_2 = i\Delta z q^{J_3+1} \tag{9.80}$$

$$T_3 = -izq^{-J_3-1} \tag{9.81}$$

and

$$T_4 = -\frac{i\Delta}{z} q^{-J_3} \tag{9.82}$$

where

$$J_3 = z\frac{\partial}{\partial z} \equiv S_3 + j \tag{9.83}$$

is the dilation operator, so that

$$q^{J_3}\psi(z) = \psi(qz) \quad . \tag{9.84}$$

Similar relationships remain valid with respect to (9.15). Choosing for convenience $\theta_1 = \theta_2 = 0$ and $\Delta = 1$, one has the Hamiltonian decomposition

$$\mathcal{H} = T_+ + T_- + z + \frac{1}{z} \tag{9.85}$$

where

$$T_\pm = q^{\pm 2J_3} \tag{9.86}$$

in which case

$$T_\pm z = q^{\pm 2} z T_\pm \tag{9.87}$$

and

$$T_\pm \frac{1}{z} = \frac{q^{\mp 2}}{z} T_\pm \quad . \tag{9.88}$$

In general, a finite magnetic translation by a vector $\vec{a}$ is produced by an operator like

$$T_{\vec{k}}(\vec{a}) = \exp i\left(\vec{k}_{op} \cdot \vec{a}\right) \tag{9.89}$$

where $\vec{k}_{op} = \left(k_1^{(op)}, k_2^{(op)}\right)$ is the wavevector operator established in accord with (8.4).

Choosing now the isotropic gauge

$$\vec{A} = \frac{B}{2}(-y, x, 0) \quad , \tag{9.90}$$

for which

$$A_i = -\frac{1}{2}B\varepsilon_{ij}x_j \tag{9.91}$$

one obtains the k_j operators as

$$k_j \rightarrow k_j^{(op)} = -i\frac{\partial}{\partial x_j} - \frac{eB}{2\hbar c}\varepsilon_{jk}x_k \tag{9.92}$$

where the non-zero elements of ε_{ij} are $\varepsilon_{12} = 1$ and $\varepsilon_{21} = -1$. Locating $\vec{a}$ and $\vec{b}$ vectors in the xy-plane, one obtains the composition law of the group of magnetic translations as

$$T_{\vec{k}}(\vec{a})\, T_{\vec{k}}\left(\vec{b}\right) = qT_{\vec{k}}\left(\vec{a} + \vec{b}\right) \tag{9.93}$$

where

$$q = \exp\left(\frac{ie}{2\hbar c}\left(\vec{a} \times \vec{b}\right) \cdot \vec{B}\right) \quad . \tag{9.94}$$

This leads to the ray representation (see also Fradkin (1991))

$$T_{\vec{k}}\left(\vec{b}\right) T_{\vec{k}}(\vec{a}) = \frac{1}{q^2} T_{\vec{k}}(\vec{a})\, T_{\vec{k}}\left(\vec{b}\right) \tag{9.95}$$

which proceeds in accord with (9.77) and (9.78).

9.6 The $SU_q(2)$-symmetry of the usual Harper Hamiltonian

The Hamiltonian characterizing the usual Harper-equation can be expressed as

$$\mathcal{H} = \frac{1}{2}\left[T_{\vec{k}}(\vec{a}) + T_{\vec{k}}(-\vec{a}) + T_{\vec{k}}\left(\vec{b}\right) + T_{\vec{k}}\left(-\vec{b}\right)\right] \tag{9.96}$$

in accord with (8.5), where $\vec{a} = (a, 0, 0)$ and $\vec{b} = (0, a, 0)$. There is $T_{\vec{k}}^{+}(\vec{a}) = T_{\vec{k}}(-\vec{a})$ under Hermitian conjugation and similarly for

$T_{\vec{k}}\left(\vec{b}\right)$. Next let us introduce $J_{\pm}$- and J_3-generators by virtue of the realizations (Alavi and Rouhani (2004))

$$J_{+} = \frac{1}{q-q^{-1}}\left(T_{\vec{k}}\left(\vec{a}\right) + T_{\vec{k}}\left(\vec{b}\right)\right) \tag{9.97}$$

$$J_{-} = J_{+}^{+} = -\frac{1}{q-q^{-1}}\left(T_{\vec{k}}\left(-\vec{a}\right) + T_{\vec{k}}\left(-\vec{b}\right)\right) \tag{9.98}$$

and

$$q^{2J_3} = T_{\vec{k}}\left(\vec{a} - \vec{b}\right) \tag{9.99}$$

respectively. It can be easily verified that

$$[J_{+}, J_{-}] = [2J_3]_q \tag{9.100}$$

by virtue of (9.93). In addition there is

$$q^{2J_3} J_{\pm} q^{-2J_3} = q^{\pm 2} J_{\pm} \tag{9.101}$$

which is synonymous with the usual relationship

$$[J_3, J_{\pm}] = \pm J_{\pm} \quad . \tag{9.102}$$

So it is clear that (9.100) and (9.102) are responsible for a typical $SU_q(2)$-symmetry. Accordingly, the Harper-Hamiltonian gets expressed as

$$\mathcal{H} = \frac{q-q^{-1}}{2}\left(J_{+} - J_{-}\right) = i(q-q^{-1})J_y \tag{9.103}$$

where we have assumed that $J_{\pm} = J_x \pm iJ_y$, as usual. Note that this $SU_q(2)$-symmetry can be viewed as a particular case of the $SL_q(2)$-realization characterizing the q-symmetrized Harper-equation.

9.7 Commutation relations concerning magnetic translation operators and the Hamiltonian

Equation (9.95) shows that the magnetic translation operators do not commute with each other in so far as $q^2 \neq 1$, as one might expect. However, we can introduce commuting operators by resorting to a rescaled primitive cell like $(\widetilde{a}, b)$ instead of (a, b). Accordingly, the flux ϕ gets replaced by

$$\widetilde{\phi} = \frac{\widetilde{a}}{a}\phi = \frac{\widetilde{a}}{a}\frac{P}{Q} \tag{9.104}$$

so that

$$q^2 \rightarrow \widetilde{q}^2 = \exp\left(2\pi i \frac{\widetilde{a}}{a}\frac{P}{Q}\right) \tag{9.105}$$

which proceeds by virtue of (7.4). Choosing

$$\widetilde{a} = aQ \quad , \tag{9.106}$$

then gives $\widetilde{q}^2 = 1$, so that the corresponding magnetic translation operators commute with each other as

$$\left[T_{\vec{k}}\left(\vec{b}\right), T_{\vec{k}}\left(Q\vec{a}\right)\right] = 0 \quad . \tag{9.107}$$

One should also have

$$k_1\widetilde{a} \in [0, 2\pi] \tag{9.108}$$

which produces in turn the "magnetic" Brillouin zone

$$k_1 a \in [0, 2\pi/Q] \quad . \tag{9.109}$$

Starting from (9.107), we then have to realize that the magnetic translation operators commute with the Hamiltonian

$$\left[\mathcal{H}, T_{\vec{k}}\left(\vec{b}\right)\right] = \left[\mathcal{H}, T_{\vec{k}}\left(Q\vec{a}\right)\right] = 0 \quad . \tag{9.110}$$

Indeed, choosing e.g. the Landau-gauge leads to the concrete realizations

$$T_1(a) = \exp\left(a\frac{\partial}{\partial x_1}\right) \tag{9.111}$$

and

$$T_2(b) = \exp\left(b\left(\frac{\partial}{\partial x_2} + \frac{ie}{\hbar c}Bx_1\right)\right) \quad . \tag{9.112}$$

Now it is an easy matter to verify that $T_1(Qa)$, $T_2(b)$ and the Hamiltonian (8.5) are mutually commuting operators, which supports in turn (9.107) and (9.110).

Chapter 10

Quantum Oscillations and Interference Effects in Nanodevices

The present miniaturization of quantum nanoscale systems like rings, quantum wires and chains of quantum dots has reached a stage where the sample dimensions are smaller than the coherence length characterizing the single electron wavefunction (Akkermans and Montambaux, 2004). The phase coherence implemented in this manner leads to interesting interference effects which are able to be verified from the experimental point of view. A typical example is provided by the oscillations characterizing the flux dependence of persistent currents. Such oscillations can be traced back to the Aharonov-Bohm effect and the same concerns the magnetoresistance oscillations in mesoscopic rings. On the other hand the application of external fields results in controllable modification of the phase coherence of electronic wavefunctions, which produce in turn quantum interference phenomena affecting specifically the electron transport.

One shows that applying Fourier series to the derivation of the total current in the discretized AB ring leads to the appearance of nontrivial odd-even parity effects. We shall then pay a special attention to rings attached to leads (Xiong and Liang (2004)), quantum wire connected to quantum dots (Orellana et al (2003a)), multichain nanorings (Chen et al (1997)), quantum LC circuits (Apenko (1989), Chen et al (2002), Flores (1995)) and last but not at least to double quantum dot systems attached to leads (Orellana et al (2002)). Such junctions serve as promising prototypes to the design of further nanodevices. Accordingly, the derivation of transport properties, and especially of he conductance (Büttiker (1985), Landauer and Büttiker (1985)), is of interest for potential technological applications in microelectronics. To this aim rather transparent descriptions will be presented. Proceeding in this manner we learn how to deal with coupled second order discrete equations, too.

10.1 The derivation of generalized formulae to the total persistent current in terms of Fourier-series

Equations (8.129) and (8.130) presented before serve to the derivation of a generalized formula for the total current in the discretized AB ring by resorting to sinusoidal Fourier-series like (Papp et al (2006))

$$f_{FS}^{(l)}(\beta) = \sum_{n=1}^{\infty} b_n \sin \frac{n\pi}{l}\beta \quad . \tag{10.1}$$

This relies on basic functions, say $f_-(\beta) = f_-(-\beta)$ and $f_+(\beta)$, defined on appropriate input intervals like $\beta \in [-l, l]$ and $\beta \in [0, l]$, respectively. It is understood that the l-parameters characterizing such basic β intervals must not be at all the same. Accordingly

$$b_n = b_n^{(-)} = \frac{1}{l} \int_{-l}^{l} f_-(\beta) \sin \frac{n\pi}{l}\beta d\beta \quad , \tag{10.2}$$

and

$$b_n = b_n^{(+)} = \frac{2}{l} \int_{0}^{l} f_+(\beta) \sin \frac{n\pi}{l}\beta d\beta \quad . \tag{10.3}$$

The periodicity condition reads

$$f_{FS}^{(l)}(\beta + 2l) = f_{FS}^{(l)}(\beta) \quad , \tag{10.4}$$

in accord with (10.1), which works in terms of selected l realizations. So one has $l = l_- = 1$, which corresponds to (8.129). One remarks immediately that one has the unit period $T_- = 2l_- = 1$ in the first case, whereas a period doubling like $T_+ = 2l_+ = 1$ occurs when $l_+ = 1$.

Applying such Fourier series yields the generalized total persistent current

$$I_g^{(-)}(\beta, N_e, N_s) = -I_F f_{FS}^{(1/2)}(\beta) G^{(-)} \tag{10.5}$$

when the number of electrons is odd, where

$$f_{FS}^{(1/2)}(\beta) = -\frac{2}{\pi} \sin\left(\frac{\pi}{N_s}\right) \sum_{n=1}^{\infty} \frac{(-1)^n n}{n^2 - 1/N_s^2} \sin(2n\pi\beta) \tag{10.6}$$

stems from

$$f_-(\beta) = \sin(2\pi\beta/N_s) \tag{10.7}$$

in (8.130), while

$$G^{(-)} = \frac{N_s}{\pi N_e}\left[1 + \cos\frac{\pi}{N_s}(N_e+1) + \right. \tag{10.8}$$

$$\left. + 2\frac{\cos(\pi(N_e+1)/2N_s)\cos(\pi(N_e-1)/2N_s)}{\sin(\pi/N_s)}\right]$$

comes from the complementary factor in (8.130).

Starting from (8.129) and proceeding in a similar manner produces the generalized total persistent current

$$I_g^{(+)}(\beta, N_e, N_s) = -I_F[F_1 g_{FS}^{(1)}(\beta) + F_2 \widetilde{g}_{FS}^{(1)}(\beta)] \tag{10.9}$$

for even values of N_e, where

$$F_1 = \frac{N_s}{\pi N_e}\left[\sin\left(\pi\frac{N_e}{N_s}\right)\cot\left(\frac{\pi}{N_s}\right) + \cos\left(\frac{\pi}{N_s}\right)\cos\left(\frac{\pi(N_E+1)}{2N_s}\right)\right] \tag{10.10}$$

and

$$F_2 = \frac{N_s}{\pi N_e}\cos\left(\frac{\pi}{N_s}\right)\sin\left(\frac{\pi(N_E+1)}{2N_s}\right) \quad . \tag{10.11}$$

This time $g_{FS}^{(1)}(\beta)$ and $\widetilde{g}_{FS}^{(1)}(\beta)$ are generated by

$$f_+(\beta) = \sin(2\pi\beta/N_s) \quad , \tag{10.12}$$

and

$$f_+(\beta) = \sin(\pi\,(2\beta - 2N_{l-1})/N_s) \quad , \tag{10.13}$$

respectively. Accordingly, one has

$$g_{FS}^{(1)}(\beta) = -\frac{2}{\pi}\sin\left(\frac{2\pi}{N_s}\right)\sum_{n=1}^{\infty}\frac{(-1)^n n}{n^2 - 4/N_s^2}\sin(n\pi\beta) \tag{10.14}$$

and

$$\widetilde{g}_{FS}^{(1)}(\beta) = -\frac{2}{\pi}\sum_{n=1}^{\infty}\left(1-(-1)^n \cos\frac{2\pi}{N_s}\right)\frac{n}{n^2-4/N_s^2}\sin(n\pi\beta) \quad . \qquad (10.15)$$

Equations (10.6), (10.14) and (10.15) have been established in terms of integrals like

$$\int \sin(ax)\cos(bx)dx = -\frac{\cos(a+b)x}{2(a+b)} - \frac{\cos(a-b)x}{2(a-b)} \qquad (10.16)$$

and

$$\int \sin(ax)\sin(bx)dx = \frac{\sin(a-b)x}{2(a-b)} - \frac{\sin(a+b)x}{2(a+b)} \qquad (10.17)$$

The flux dependence of dimensionless total persistent currents like

$$C = C_g^{(\pm)}(\beta, N_e, N_s) = \frac{I_g^{(\pm)}(\beta, N_e, N_s)}{I_F} \qquad (10.18)$$

is displayed in figures 10.1 and 10.2 for $N_e = 3$ and $N_e = 4$, respectively, where $N_s = 12$.
Such results stand for the discrete counterpart of (2.122). In other words we succeeded to establish a period doubling as well as two kinds of distinct amplitudes when passing, in a way or another, from odd values of the electron number to even ones and conversely. Besides the dynamic localization effects discussed in chapter 6, such effects rely intimately on the space discreteness, which also means that they would disappear within the continuous limit. We found nontrivial odd-even parity effects in the flux-dependence of the total current, which concerns both the period of oscillations as well as the corresponding magnitudes. The period referred to above can be readily expressed as

$$T_\Phi^{(N_e)} = \left[\frac{1}{2} + \frac{1+(-1)^{N_e}}{4}\right] 2\Phi_0 \quad , \qquad (10.19)$$

where N_e may be even or odd. There is still a point which remains open for further investigations, namely the identification of nanoring-structures reproducing period doubling effects presented above form the experimental point of view. The influence of disorder and interaction on the persistent currents has also been discussed (Fye et al (1991), Németh and Pichard (2005), Carvalo Dias et al (2006)).

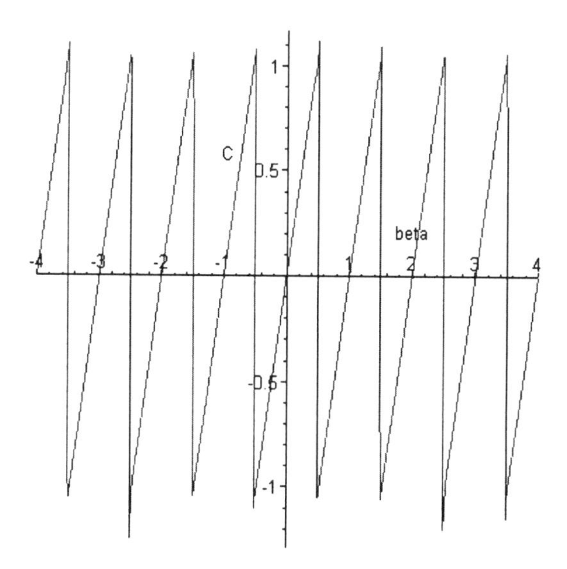

Fig. 10.1 **The flux dependence of** $C_g^{(-)}(\beta, N_e, N_s)$ **for** $N_e = 3$ **and** $N_s = 12$. **The current oscillations displayed above exhibit the unit period, while the amplitudes have the magnitude order of unity.**

10.2 The discretized Aharonov-Bohm ring with attached leads

A further configuration is done by discretized Aharonov-Bohm ring with N sites, now with two attached semi-infinite leads. These leads are attached to the sites 1 and n, as shown in figure 10.3

The point-like couplings between the ring and the leads are characterized by the hopping amplitudes t_R and t_L, where the subscripts "R" and "L" stand for "right" and "left", respectively. A pointable coupling between the leads expressed by the t_c will also be accounted for. This latter coupling provides the continuous path for inter-lead electron transmission. In addition, there are tunneling effects of the electrons through the ring. In order to proceed further let us denote the creation (annihilation) operator of the spinless electron on the ring, or the left and right leads by c_l^+ (c_l), a_m^+ (a_m) and b_m^+ (b_m), respectively. The corresponding site-numbers are given by $l = 1, 2, \ldots$, $m = -1, -2, \ldots$ for the left-lead, whereas $m = 1, 2, \ldots$ for the right lead. In order to describe the non-interacting leads, i.e. the pertinent 1D conductors, we shall resort to 1D tight-binding models with

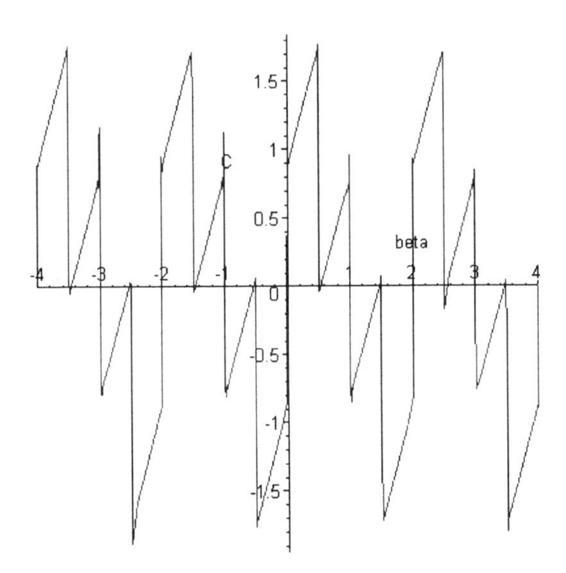

Fig. 10.2 **The flux dependence of $C_g^{(+)}(\beta, N_e, N_s)$ for $N_e = 4$ and $N_s = 12$. This time the current oscillations exhibit the double period. In addition, there are two kinds of amplitudes for which the magnitude orders are given nearly by 1.74 and 0.84.**

NN in interaction. This gives the Hamiltonian

$$\mathcal{H}_{leads} = t_0 \sum_{m\leqslant -1} \left(a_m^+ a_{m+1} + a_{m+1}^+ a_m\right) + t_0 \sum_{m\geqslant 1} \left(b_m^+ b_{m+1} + b_{m+1}^+ b_m\right) \quad , \tag{10.20}$$

where t_0 stands for the inherent hopping parameter.

The electron on the ring is described by the Hamiltonian

$$\mathcal{H}_{Ring} = \sum_{l=1}^{n} \left(\varepsilon_l c_l^+ c_l + t_r \exp\left(i\varphi_R\right) c_l^+ c_{l+1} + t_r \exp\left(-i\varphi_R\right) c_{l+1}^+ c_l\right) \tag{10.21}$$

where the site energy is given by ε_l and where t_R denotes the related hopping parameter. One realizes that (8.139) proceeds in a close analogy with (7.1). Accordingly, the corresponding phase reads

$$\varphi_R = \frac{2\pi}{N}\frac{\phi}{\phi_0} = \frac{2\pi}{N}\beta \quad , \tag{10.22}$$

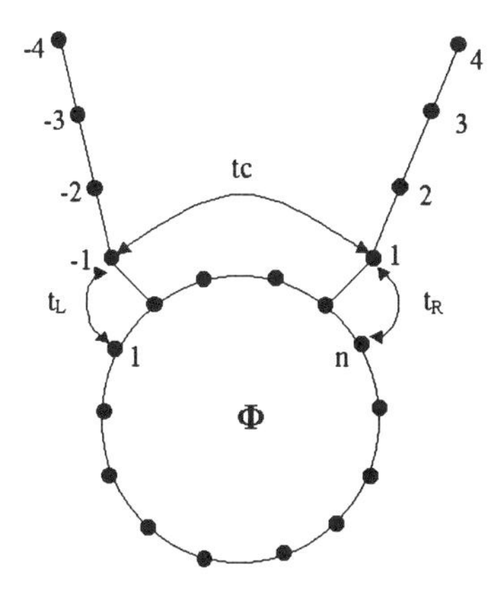

Fig. 10.3 **Schematic view of the quantum ring threaded by magnetic flux and attached to two leads.**

which is reminiscent to (7.3). The periodic boundary condition characterizing (10.21) reads $c_{l+n} = c_l$. Putting together both point-like transmission channels leads to the tunneling Hamiltonian (Xiong and Liang (2004))

$$\mathcal{H}_T = t_c a_{-1}^+ b_1 + t_L a_{-1}^+ c_1 + t_R b_1^+ c_n + \qquad (10.23)$$

$$+ t_c b_1^+ a_{-1} + t_L c_1^+ a_{-1} + t_R b_1 c_n^+$$

which is responsible for the transport properties one looks for.

At this point we have to realize that a uniform gate voltage, say V_g, can be introduced via

$$\varepsilon_l = \varepsilon_l^0 + V_g \qquad (10.24)$$

which provides the tuning-parameter for further investigations. It can be assumed, for convenience, that the zero-voltage energy is zero, i.e. that $\varepsilon_l^0 = 0$. It is understood that the spinless electron operators mentioned above satisfy usual canonical commutations relations like $[a_m, a_n^+] = \delta_{m,n}$ and similarly for b_m and c_l. In addition one has

$$[a_m, b_n] = [a_m, c_l] = [b_n, c_l] = 0 \quad . \qquad (10.25)$$

At the beginning we shall discuss the solution of the time dependent Schrödinger equation

$$\mathcal{H}\left|\psi\left(t\right)\right\rangle = i\hbar\frac{\partial}{\partial t}\left|\psi\left(t\right)\right\rangle \quad , \tag{10.26}$$

where the total Hamiltonian reads

$$\mathcal{H} = \mathcal{H}_{Leads} + \mathcal{H}_{Ring} + \mathcal{H}_{T} \tag{10.27}$$

Correspondingly, we shall choose the wavefunction as

$$|\psi(t)\rangle = \sum_{m\leq -1} A_m(t)a_m^+ |0\rangle + \sum_{m\geq 1} B_m(t)b_m^+ |0\rangle + \sum_{l=1}^{n} C_l(t)c_l^+ |0\rangle \, , \tag{10.28}$$

where we have assumed that $A_0 = B_0 = 0$. The corresponding time dependent amplitudes are denoted by $A_m(t)$, $C_l(t)$ and $B_m(t)$, respectively. This yields the coupled equations

$$i\hbar\frac{\partial}{\partial t}A_m(t) = t_0\left(A_{m+1}(t) + A_{m-1}(t)\right) + \left(t_c B_1(t) + t_L C_1(t)\right)\delta_{m,-1} \quad , \tag{10.29}$$

for $m \leq -1$,

$$i\hbar\frac{\partial}{\partial t}B_m(t) = t_0\left(B_{m+1}(t) + B_{m-1}(t)\right) + \left(t_R C_n(t) + t_c A_{-1}(t)\right)\delta_{m,1} \tag{10.30}$$

for $m \geq 1$, and

$$\left(i\hbar\frac{\partial}{\partial t} - \varepsilon\right)C_l(t) = t_r C_{l+1}(t)\exp\left(i\varphi_R\right) + t_r C_{l-1}(t)\exp\left(-i\varphi_R\right) + \tag{10.31}$$
$$+ t_L A_{-1}(t)\delta_{l,1} + t_R B_1(t)\delta_{l,n} \quad ,$$

which are responsible for typical manifestations. The above equations correspond to the stationary solutions which have been written down before by Xiong and Liang (2004). These latter solutions can be readily obtained by inserting

$$|\psi(t)\rangle = \exp\left(-iE\frac{t}{\hbar}\right)|\psi\rangle \tag{10.32}$$

into (10.26), where $|\psi\rangle = |\psi(0)\rangle$ and $\mathcal{H}\,|\psi\rangle = E\,|\psi\rangle$. Accordingly there is

$$C_l(t) = \exp\left(-iE\frac{t}{\hbar}\right) C_l \tag{10.33}$$

where $C_l(0) = C_l$. The same concerns A_m and B_m.

Next let us multiply (10.31) by $C_l^*(t)$. Repeating the usual procedure to the quantum-mechanical derivation of currents leads to

$$\frac{\partial}{\partial t} \mid C_l(t) \mid^2 = \Delta I_l + \frac{2}{\hbar} Im\left(t_L A_{-1}(t) C_1^*(t)\delta_{l,1} + t_R B_1(t) C_n^*(t)\delta_{l,n}\right) \quad , \tag{10.34}$$

where

$$I_l = \frac{2}{\hbar} t_R Im\left(C_{l-1}^*(t) C_l(t) \exp(i\varphi_R)\right) \quad . \tag{10.35}$$

Note that Δ stands for the discrete right-hand derivative introduced before in accord with (1.3). So far we are able to introduce the electric charge density as

$$\rho_l(t) = -e \mid C_l(t) \mid^2 \tag{10.36}$$

in which case the current density reads

$$J_l = eI_l \tag{10.37}$$

in so far as $l \neq 1$ and $l \neq n$.

We have to realize that the additional terms characterizing (10.34) can be interpreted in terms of rate equations like

$$\left(\frac{d}{dt}\rho_1(t)\right)_{Ring-Leads} = -\frac{2e}{\hbar} t_L Im\left(A_{-1}(t) C_1^*(t)\right) \tag{10.38}$$

and

$$\left(\frac{d}{dt}\rho_n(t)\right)_{Ring-Leads} = -\frac{2e}{\hbar} t_R Im\left(B_1(t) C_n^*(t)\right) \quad , \tag{10.39}$$

which are responsible for the flow of electrons into and out of the ring. Charge conservation requirements needed are then fulfilled if

$$\frac{d}{dt}\left(\rho_1(t) + \rho_n(t)\right)_{Ring-Leads} = 0 \tag{10.40}$$

so that

$$Im\left(t_L A_{-1}(t) C_1^*(t) + t_R B_1(t) C_n^*(t)\right) = 0 \quad . \tag{10.41}$$

So we are in a position to establish the average ring current as

$$J = < J_l > = \frac{1}{n} \sum_{l=1}^{n} J_l \quad , \tag{10.42}$$

which proceeds in accord with charge conservation requirements just discussed above.

Incoming and outgoing interaction-field regions can be identified via $m \leq -2$ and $m \geq 2$, respectively. Within such interaction-free regions equations A_m and B_m are characterized by plane waves like

$$A_m = \exp\left(ik\left(m+1\right)\right) + r \exp\left(-ik\left(m+1\right)\right) \quad , \tag{10.43}$$

and

$$B_m = t \exp\left(ikm\right) \tag{10.44}$$

where k is the dimensionless wave number, while r and t denote reflection and transmission amplitudes. The energy of the incident electron is given by

$$E_{in} = 2t_0 \cos k \quad , \tag{10.45}$$

which reproduces identically (6.4) in terms of units for which the lattice spacing is unity. Concerning the ring, the interaction-free regions are specified by $l \neq 1$ and $l \neq n$. This gives the equation

$$\left(E - \varepsilon\right) C_l = t_r C_{l+1} \exp\left(i\varphi_R\right) + t_r C_{l-1} \exp\left(-i\varphi_R\right) \quad , \tag{10.46}$$

in accord with (10.31) and (10.32). Invoking again plane-wave solutions

$$C_l = \exp\left(i\widetilde{k}l\right) \tag{10.47}$$

enables us to derive the ring energy as

$$E = E_r = \varepsilon_l + 2t_r \cos\left(\widetilde{k} + \varphi_R\right) \quad . \tag{10.48}$$

One sees that (10.48) reproduces identically (8.122) via $\varepsilon_l = 0$, $t_r = -V_0$ and $\widetilde{k} = 2\pi n/N_s$. One would then have $\widetilde{k} \in [0, 2\pi)$ if $n \in [0, N_s)$, which means that we deals with N_s-levels ($n = 0, 1, 2, \ldots, N_s - 1$).

Of a special interest is the transmission probability

$$T_p = |B_1|^2 \quad , \tag{10.49}$$

which provides the conductance by the virtue of the well-known relationship (see e.g. Datta (1995))

$$G_0 = \frac{2e^2}{h} G = \frac{2e^2}{h} T_p \quad . \tag{10.50}$$

To this aim it can be easily verified, that

$$A_{-2} = A_{-1} \exp(ik) - 2i \sin k \tag{10.51}$$

and

$$B_2 = \exp(ik) B_1 \quad . \tag{10.52}$$

This yields the equations

$$A_{-1} \left(E - t_0 \exp(ik)\right) - t_c B_1 = -2it_0 \sin k + t_L C_1 \tag{10.53}$$

and

$$t_c A_{-1} - B_1 (E - t_0 \exp(ik)) = -t_R C_n \quad . \tag{10.54}$$

Fixing the energy, we then have the opportunity to establish A_{-1} and B_1 in terms of C_1 and C_n. One would then have

$$B_1 = \frac{it_R C_n + 2it_c - t_L t_c C_1}{1 + t_c^2} \quad , \tag{10.55}$$

where the couplings are measured in units of t_0. What then remains is to solve numerically coupled equations characterizing C_l, A_m and B_m. Such studies have been done in terms of the fixing $E = E_{in} = 0$ (Xiong and Liang (2004)). Quantum rings coupled to leads have also been discussed under the influence of an external electric field (Orellana et al (2003b)).

In general, there are reasons to say that both conductance and persistent current are periodic functions of the magnetic flux with period ϕ_0. However, a period doubling is able to occur for selected parameter values. Indeed, choosing parameters like $V_g = 0.03$, $N_s = 5$, $t_L = t_R = 0.1$ and $k = \pi/2$, we found that the period characterizing the conductance oscillations with respect to the magnetic flux is given by $2\phi_0$ instead of ϕ_0, as indicated in figure 10.4. In addition, there are peaks reproducing specifically asymmetric Fano line-shapes (Fano (1961)) in the dependence of G on the dimensionless magnetic flux β, as well as on the gate voltage V_g. Such resonance peaks, which are associated with the levels of the ring are a manifestation of quantum interference effects characterizing the electron transmission along the two paths mentioned before. Narrow resonance peaks have been observed in the dependence of the persistent current (10.41) on the gate voltage and the magnetic flux too. It is then clear that the asymmetric Fano-profiles get replaced by symmetric Breit-Wigner line shapes when $t_c \to 0$. Peculiarities concerning the related Fano-parameter have also been discussed (Kobayshi et al (2004), Xiong and Liang (2004,2005)). In particular, the present ring configuration can also be viewed as a quantum dot.

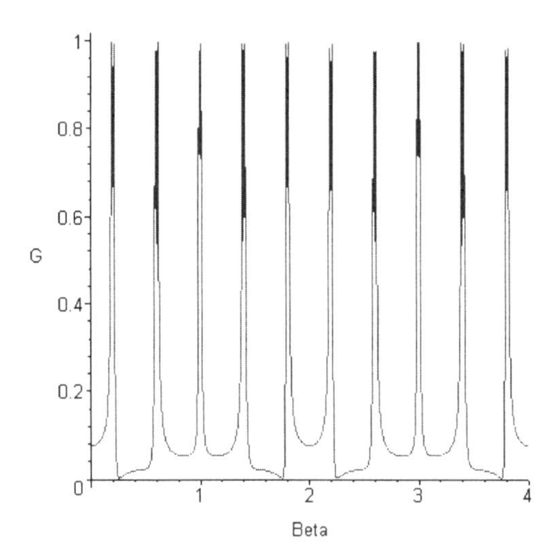

Fig. 10.4 **The oscillations of the dimensionless conductance versus β for $V_g = 0.03$, $N_s = 5$, $t_L = t_R = 0.1$ and $k = \pi/2$. The double period can be readily identified.**

Without considering other details, we have to mention that electrons confined on a quantum Aharonov-Bohm ring are able to form a spin singlet state with electrons in the leads. This results in the implementation of a pronounced many-body Kondo effect at lower fields, which received much attention during last decades (Ferrari et al (1999), Kang and Shin (2000), Keyser et al (2003), Gomez et al (2004)).

10.3 Quantum wire attached to a chain of quantum dots

A further interesting nanodevice is a side coupled to a chain of quantum dots as shown in figure 10.5. The Hamiltonian reads (Orellana et al (2003a))

$$\mathcal{H} = \mathcal{H}_{QW} + \mathcal{H}_{QD} + \mathcal{H}_{QD-QW} \quad , \tag{10.56}$$

in which $\mathcal{H}_{QW}$ ($\mathcal{H}_{QD}$) is responsible for the quantum wire (the chain of quantum dots), while $\mathcal{H}_{QD-QW}$ stands for the tunneling interaction between the quantum wire and the quantum dots. One has

$$\mathcal{H}_{QW} = V \sum_{j=-\infty}^{+\infty} \left(c_j^+ c_{j+1} + H.c. \right) \quad , \tag{10.57}$$

within the nearest neighbor description, where V denotes the hopping parameter. The electron at site j is created by c_j^+, as usual. The Hamiltonian describing the chain of N quantum dots is

$$\mathcal{H}_{QD} = \sum_{l=1}^{N} \varepsilon_l d_l^+ d_l + \sum_{l=1}^{N-1} \left(V_l d_l^+ d_{l+1} + H.c. \right) \quad , \tag{10.58}$$

where V_l is a real parameter denoting the tunneling coupling between the l-th and $(l+1)$-th quantum dots. The tunneling interaction

$$\mathcal{H}_{QD-QW} = \widetilde{U}_0 \left(d_1^+ c_0 + c_0^+ d_1 \right) \tag{10.59}$$

concerns only the electrons located at $j = 0$ and $l = 1$, respectively, as indicated in figure 10.5.

One looks for stationary states like

$$|\Psi\rangle = \sum_{j=-\infty}^{+\infty} A_j c_j^+ |0\rangle + \sum_{l=1}^{N} D_l d_l^+ |0\rangle \quad , \tag{10.60}$$

where A_j and D_l are expansion coefficients providing probability amplitudes needed. One deals again with Wannier states like $|j\rangle = c_j^+ |0\rangle$ and

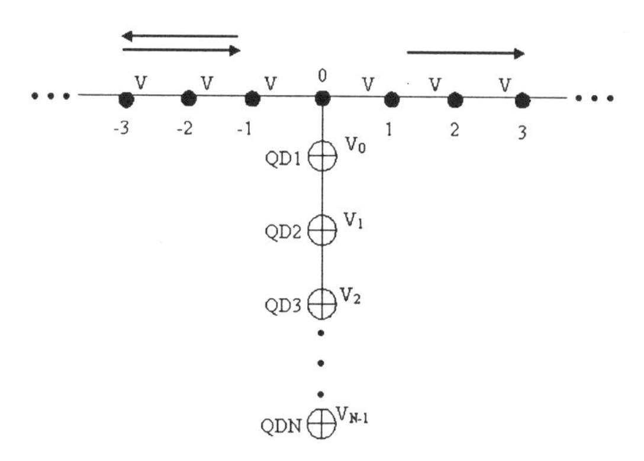

Fig. 10.5 **Schematic view of the quantum wire attached to a chain of quantum dots.**

$|l\rangle\rangle = d_l^+ |0\rangle$, such that $\langle j|\, j'\rangle = \delta_{j,j'}$ and $\langle\langle l|\, l'\rangle\rangle = \delta_{l,l'}$. Conversions to Bloch states can also be easily performed.

Now we are ready to solve again the eigenvalue equation $\mathcal{H}\,|\Psi\rangle = E\,|\Psi\rangle$ in terms of linear second-order discrete equations for A_j and D_l. These equations are given by

$$EA_0 = V\,(A_{-1} + A_1) + \widetilde{U}_0 D_1 \quad , \tag{10.61}$$

$$EA_j = V\,(A_{j-1} + A_{j+1})\,, \quad j \neq 0 \quad , \tag{10.62}$$

$$ED_N = \varepsilon_N D_N + V_{N-1} D_{N-1} \quad , \tag{10.63}$$

$$ED_1 = \varepsilon_1 D_1 + V_1 D_2 + \widetilde{U}_0 A_0 \quad , \tag{10.64}$$

and

$$ED_l = \varepsilon_l D_l + V_{l-1} D_{l-1} + V_l D_{l+1}, \quad l \neq 1, N \quad . \tag{10.65}$$

The interesting point is that D_1 can be expressed in terms of A_0 as (Orellana et al (2003a))

$$D_1 = \frac{V_0}{Q_N} A_0 \quad , \tag{10.66}$$

where Q_N is a continued function:

$$Q_N = Q_N(E) = E - \varepsilon_1 - \cfrac{V_1^2}{E - \varepsilon_2 - \cfrac{V_2^2}{E-\varepsilon_3-\ldots E-\varepsilon_{N-1}-V_{N-1}^2/(E-\varepsilon_N)}} \quad . \tag{10.67}$$

Proceeding further we then have to solve the equations

$$EA_{-1} = V(A_{-2} + A_0) \quad , \tag{10.68}$$

$$\widetilde{E}A_0 = V(A_{-1} + A_1) \quad , \tag{10.69}$$

and

$$EA_1 = V(A_0 + A_2) \quad , \tag{10.70}$$

where now

$$\widetilde{E} = E - \frac{U_0^2}{Q_N} \quad . \tag{10.71}$$

At this point, we have to realize that the energy is given by

$$E = E(k) = 2V\cos k \quad , \tag{10.72}$$

where the dimensionless wave number k is restricted to the first Brillouin zone $k \in [-\pi, \pi]$. For this purpose the wavefunction ansatz

$$A_j = \exp(ikj) + r\exp(-ikj)\,, \quad j \leq -1 \quad , \tag{10.73}$$

$$A_j = t\exp(ikj) \tag{10.74}$$

is used once more again, where r and t are reflection and transmission amplitudes, respectively. Extrapolating the above wavefunctions towards $j = 0$, one finds the matching condition

$$t - r = 1 \quad , \tag{10.75}$$

which provides an appreciable simplification. Inserting $E = E(k)$ then gives the transmission amplitude

$$t = A_0(E) = \frac{Q_N(E)}{Q_N(E) - iU_0^2/\sqrt{4v^2 - E^2}} \quad . \tag{10.76}$$

This shows that the level broadening Γ can be identified as

$$\Gamma = \Gamma(E) = \frac{U_0^2}{\sqrt{4v^2 - E^2}} \quad . \tag{10.77}$$

In other words, the conductance of the quantum wire at zero temperature reads

$$G(E) = \frac{2e^2}{h} \frac{Q_N^2}{Q_N^2 + T^2} \quad , \tag{10.78}$$

by virtue of the one-chanell Landauer-formula (Datta (1995)), where the transmission coefficient is given, this time, by

$$T_N(E) = |t|^2 = |A_0|^2 = \frac{Q_N^2}{Q_N^2 + T^2} \quad . \tag{10.79}$$

Resonance structures characterizing the energy dependence of $T_N(E)$ can then be easily identified by looking for complex $E = E_c-$roots for which $T_N(E_c) = 1$, as shown before (Orellana et al (2003a)).

10.4 Quantum oscillations in multichain nanorings

Let us remember that the magnetic field induces characteristic phase changes like

$$\arg\psi(\vec{x}) \rightarrow \arg\psi(\vec{x}) - \frac{e}{\hbar c}\int_1^2 \vec{A}\cdot\vec{dl} \quad , \tag{10.80}$$

along a path ranging from "1" to "2", where $-e < 0$ stands for the electron charge. For a closed path, the electron wavefunction experiences the phase-difference $-2\pi\Phi/\Phi_0$, as discussed before in connection with the Aharonov-Bohm boundary condition. Such phase-shifts produce interesting interference phenomena in the case of nanosystems for which the coherence length is actually larger than the sample dimension. Interference effects implemented in this manner are also of a special interest in the case of a multichain nanorings such as displayed in figure 10.6.

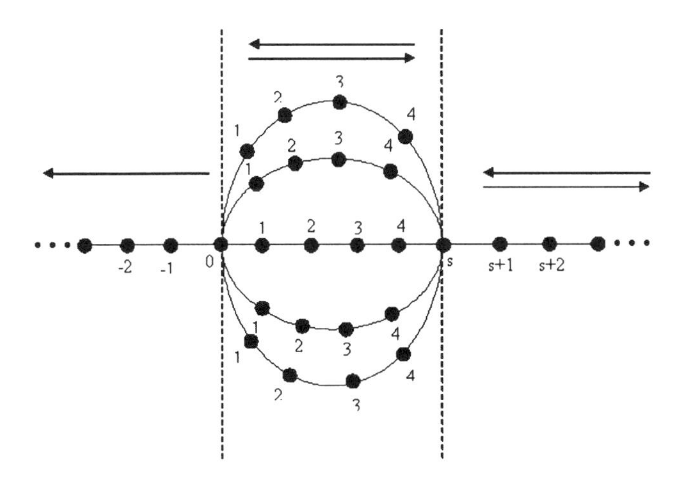

Fig. 10.6 **Schematic view of multichain nanorings. One has 5 chains, each of which containing, just for convenience, 4 sites.**

The chains are specified by the index $\alpha = 1, 2, \ldots, N$, while the number of sites characterizing a chain is $0 < N_\alpha < s$. These chains are attached to two leads at nodes located at $x = 0$ and $x = s > 0$. So one deals with the influence of the magnetic flux $\Phi_\alpha = BS_\alpha$, where S_α is the surface enclosed by chains α and $\alpha - 1$ $(\alpha = 2, 3, \ldots, N)$. In addition, we shall account for the influence of a voltage V_α applied in the transverse direction. The present Hamiltonian is then given by the sum (Chen et al (1997))

$$\mathcal{H} = \mathcal{H}_{leads} + \mathcal{H}_{chains} + \mathcal{H}_{int} \quad , \tag{10.81}$$

and concerns, as before, non-interacting electrons. The individual terms in (10.81) are responsible successively for the leads, the chains and for the tunneling interaction. Furthermore, let us consider that $c^+_{\alpha,i}$ $\left(c^+_j\right)$ creates an electron at the site i of the chain α (at the site j of the leads). Accordingly, one has

$$\mathcal{H}_{leads} = \sum_{j=-\infty}^{0} \left(c^+_j c_{j-1} + c^+_{j-1} c_j\right) + \sum_{j=s}^{\infty} \left(c^+_j c_{j+1} + c^+_{j+1} c_j\right) \quad , \tag{10.82}$$

$$\mathcal{H}_{chains} = V_\alpha \sum_{\alpha=1}^{N} \left(\sum_{i=1}^{N_\alpha} c^+_{\alpha,i} c_{\alpha,i} + \sum_{i=1}^{N_\alpha - 1} \left(c^+_{\alpha,i} c_{\alpha,i+1} + c^+_{\alpha,i+1} c_{\alpha,i}\right) \right) \quad , \tag{10.83}$$

and

$$\mathcal{H}_{int} = \sum_{i=1}^{N} \left(c_0^+ c_{\alpha,1} + c_{\alpha,1}^+ c_0 + \exp\left(i\varphi_\alpha\right) c_{\alpha,N_\alpha}^+ c_s + \exp\left(-i\varphi_\alpha\right) c_s^+ c_{\alpha,N_\alpha} \right) \quad . \tag{10.84}$$

For convenience a unit hopping amplitude has been assumed. The present gauge concerns only the site $i = N_\alpha$, such that

$$2\pi \frac{\Phi_\alpha}{\Phi_0} = \varphi_\alpha - \varphi_{\alpha-1} \quad , \tag{10.85}$$

for $\alpha \geq 2$, where $\varphi_1 = 0$.

Now we are ready to handle the energy eigenvalue problem $\mathcal{H}\left|\Psi\right\rangle = E\left|\Psi\right\rangle$, where this time

$$\left|\Psi\right\rangle = \sum_{j \leq 0, j \geq s} A_j c_j^+ \left|0\right\rangle + \sum_{\alpha=1}^{N} \sum_{i=1}^{N_\alpha} A_{\alpha,i} c_{\alpha,i}^+ \left|0\right\rangle \quad . \tag{10.86}$$

These results again in coupled discrete equations like

$$E A_j = A_{j+1} + A_{j-1} \tag{10.87}$$

for $j \leq -1$ and $j \geq s+1$,

$$E A_0 = \sum_\alpha A_{\alpha,1} + A_{-1} \quad , \tag{10.88}$$

$$E A_s = \sum_\alpha \exp\left(-i\varphi_\alpha\right) A_{\alpha,N_\alpha} + A_{s+1} \quad , \tag{10.89}$$

$$\left(E - V_\alpha\right) A_{\alpha,1} = A_0 + A_{\alpha,2} \quad , \tag{10.90}$$

$$\left(E - V_\alpha\right) A_{\alpha,N_\alpha} = A_{\alpha,N_\alpha-1} + \exp\left(i\varphi_\alpha\right) A_s \quad , \tag{10.91}$$

and

$$\left(E - V_\alpha\right) A_{\alpha,i} = A_{\alpha,i+1} + A_{\alpha,i-1} \quad , \tag{10.92}$$

where $2 \leq i \leq N_\alpha - 1$. Extrapolating (10.92) towards $i = 1$ and $i = N_\alpha$ yields the matching conditions

$$A_0 = A_{\alpha,0} \quad , \tag{10.93}$$

and

$$A_s = \exp\left(-i\varphi_\alpha\right) A_{\alpha,N_\alpha+1} \quad , \tag{10.94}$$

in accord with (10.90) and (10.91).

At this point we have to resort again to plane waves:

$$A_j = A \exp\left(-ik\left(j-s\right)\right) + R \exp\left(ik\left(j-s\right)\right) \quad , \tag{10.95}$$

and

$$A_j = \exp\left(-ikj\right) \quad , \tag{10.96}$$

for $j \geq s$ and $j \leq 0$, respectively. This means that one deals with a reflected (transmitted)) wave if $j \geq s$ ($j \leq 0$). In addition, we shall account for the superposition

$$A_{\alpha,l} = A_\alpha \exp\left(ik_\alpha l\right) + R_\alpha \exp\left(-ik_\alpha l\right) \quad , \tag{10.97}$$

acting for $0 \leq l \leq N_\alpha$. The corresponding wave numbers are identified by virtue of the energy solutions

$$E = 2\cos k \quad , \tag{10.98}$$

and

$$E = V_\alpha + 2\cos k_\alpha \quad , \tag{10.99}$$

relying on (10.87) and (10.92).

Accordingly, there is

$$A_\alpha + R_\alpha = 1 \quad , \tag{10.100}$$

and

$$A_\alpha \exp\left(ik_\alpha\left(N_\alpha+1\right)\right) + R_\alpha \exp\left(-ik_\alpha\left(N_\alpha+1\right)\right) = \left(A+R\right)\exp\left(i\varphi_\alpha\right), \tag{10.101}$$

which produce in turn the solutions

$$A_\alpha = -\frac{\exp\left(-ik_\alpha\left(N_\alpha+1\right)\right)-\left(A+R\right)\exp\left(i\varphi_\alpha\right)}{2i\sin\left(k_\alpha\left(N_\alpha+1\right)\right)} \quad , \tag{10.102}$$

and

$$R_\alpha = \frac{\exp\left(ik_\alpha\left(N_\alpha+1\right)\right)-\left(A+R\right)\exp\left(i\varphi_\alpha\right)}{2i\sin\left(k_\alpha\left(N_\alpha+1\right)\right)} \quad . \tag{10.103}$$

Under such conditions one obtains

$$E = \exp\left(ik\right)+\sum_{\alpha=1}^{N}\left(A_\alpha\exp\left(ik_\alpha\right)+R_\alpha\exp\left(-ik_\alpha\right)\right) \quad , \tag{10.104}$$

and

$$E\left(A+R\right) = A\exp\left(-ik\right)+R\exp\left(ik\right)+ \tag{10.105}$$

$$+\sum_{\alpha=1}^{N}\exp\left(-i\varphi_\alpha\right)\left(A_\alpha\exp\left(ik_\alpha N_\alpha\right)+R_\alpha\exp\left(-ik_\alpha N_\alpha\right)\right) \quad ,$$

respectively. These equations can be rewritten equivalently as

$$E = c_0+\left(A+R\right)f_0+\exp\left(ik\right) \tag{10.106}$$

and

$$E = \left(A+R\right)c_0+f_0^*+A\exp\left(-ik\right)+R\ \exp\left(ik\right) \quad , \tag{10.107}$$

where

$$c_0 = \sum_\alpha\frac{\sin\left(k_\alpha N_\alpha\right)}{\sin k_\alpha\left(N_\alpha+1\right)} \quad , \tag{10.108}$$

and

$$f_0 = \sum_\alpha\frac{\sin k_\alpha\exp\left(i\varphi_\alpha\right)}{\sin k_\alpha\left(N_\alpha+1\right)} \quad . \tag{10.109}$$

Now we are in a position to establish the A parameter as

$$A = \frac{|f_0|^2 - (c_0 - \exp(-ik))^2}{2 f_0 \cos k} \quad , \tag{10.110}$$

which provides in turn the transmission probability

$$|t|^2 = \frac{1}{|A|^2} = \frac{4 \cos^2 k \, |f_0|^2}{\left| |f_0|^2 - (c_0 - \exp(-ik))^2 \right|^2} \quad . \tag{10.111}$$

In particular, we can choose chains with the same length $N_1 = N_2 = \ldots = N_N = L$, for which the applied voltage is zero. Then we can insert $\varphi_\alpha = 2\pi(\alpha - 1)\beta$ and $\Phi_\alpha = \Phi \equiv \beta\Phi_0$ for $\alpha \geq 2$. Accordingly, f_0 becomes

$$f_0 = f_0(\beta) = \frac{\sin k \sin(\pi N \beta)}{\sin k (L+1) \sin(\pi\beta)} \exp(i\pi(N-1)\beta) \quad , \tag{10.112}$$

where now $k_\alpha = k$, so that $f_0(\beta + 2) = f_0(\beta)$ irrespective of N. In other words the periodicity characterizing the magnetic flux dependence of f_0 is give by a double flux quantum Φ_0. The rich oscillations structures exhibited by the flux and voltage dependencies of (10.111) have been presented in some detail before (Chen et al (1997)). Having established the transmission probability, enables us to study the electronic transport by applying Landauer-type conductance formulae. Other details concerning multiring systems connected in parallel and in series are worthy of being mentioned (Liu et al (1998)).

10.5 Quantum LC-circuits with a time-dependent external source

Besides the space-discretization which has been analyzed in some detail previously, other discretizations, such as the one of the electric charge in LC-circuits, has attracted much interest (Apenko (1989), Chen et al (2002), Flores (2005)). The total energy characterizing the classical LC-circuit is given by

$$\mathcal{H}_{cl} = \frac{\Phi_{cl}^2}{2Lc^2} + \frac{Q_{cl}^2}{2C} - Q_{cl} V_s(t) \quad , \tag{10.113}$$

which proceeds under the influence of an external time-dependent voltage source $V_s(t)$. The inductance and the capacitance are denoted by L and C, as usual. The subscript "cl" stands for "classical". The charge discreteness referred to above can be implemented by virtue of the eigenvalue-equation

$$Q|m\rangle = mq_e|m\rangle \quad , \tag{10.114}$$

where m is an integer and where q_e denotes the fundamental electric charge, say $q_e = e$, or eventually, $q_e = 2e$. Next let us apply to (10.114) the right-hand and left-hand discrete derivatives Δ and ∇ presented before in (1.4) and (1.3), respectively. One obtains $\Delta|m\rangle = |m+1\rangle - |m\rangle$ and $\nabla|m\rangle = |m\rangle - |m-1\rangle$, so that

$$Q\Delta = q_e + q_e(m+1)\Delta \quad , \tag{10.115}$$

and

$$Q\nabla = q_e + q_e(m-1)\nabla \quad . \tag{10.116}$$

Accordingly

$$[\Delta, Q] = -q_e(1+\Delta) \quad , \tag{10.117}$$

and

$$[\nabla, Q] = -q_e(1-\nabla) \quad , \tag{10.118}$$

by virtue of the Hermitian conjugation of (10.115) and (10.116), where $\Delta^+ = -\nabla$ and $Q^+ = Q$.

On the other hand, $Q^2/2C$ leads to an harmonic oscillator by virtue of (10.114), which also means that

$$\mathcal{H}_0 = \frac{1}{2Lc^2}\Phi_0^2 \tag{10.119}$$

plays the role of the kinetic-energy. It is understood that the square flux-operator Φ_0^2 should be established in an appropriate manner in terms of discrete derivatives displayed above. To this aim we have to realize that $\mathcal{H}_0$ relies on Δ and ∇ via

$$\mathcal{H}_0 \sim \Delta\nabla = \Delta - \nabla \quad , \tag{10.120}$$

as indicated e.g. by (1.5). Next we have to select the proportionality factor needed, which results in the tight binding realization

$$\mathcal{H}_0 = -\frac{\hbar^2}{2Lq_e^2}\Delta\nabla \quad . \tag{10.121}$$

Now we have the opportunity to introduce the magnetic flux operator

$$\Phi = -\frac{i\hbar c}{q_e}\Delta \quad , \tag{10.122}$$

which behaves as

$$\Phi^+ = -\frac{i\hbar c}{q_e}\nabla \quad , \tag{10.123}$$

under Hermitian conjugation. The Hermitian square flux operator can then be readily established as

$$\Phi_0^2 = \left(\Phi_0^2\right)^+ = \Phi\Phi^+ = \Phi^+\Phi \quad . \tag{10.124}$$

Both Φ and Φ^+ have the meaning of momentum operators conjugated to Q. A symmetrized momentum operator like

$$P = P^+ = \frac{1}{2}(\Phi + \Phi^+) \quad , \tag{10.125}$$

can also be introduced. This results in the commutation relations

$$[Q, P] = -i\hbar c\left(1 - \frac{q_e^2 L}{\hbar^2}\mathcal{H}_0\right) \quad , \tag{10.126}$$

relying on (1.62) and

$$[\mathcal{H}_0 L, Q] = i\hbar P/c \quad . \tag{10.127}$$

Modified commutation relations such as given by (10.126) have also been used before in connection with the q-deformation of quantum mechanics (Kempf et al (1995), Nouicer (2006)). Such modified structures, which are responsible for deformed Heisenberg algebras, can be viewed again as nontrivial manifestations of the space discreteness. It is understood that now the discretized charge plays the role of the discrete space.

Resorting to the m-representation

$$C_m(t) = < m \mid \Psi(t) > \quad , \tag{10.128}$$

and applying the time-dependent Schrödinger Hamiltonian

$$\mathcal{H} = -\frac{\hbar^2}{2Lq_e^2}\Delta\nabla + \frac{m^2 q_e^2}{2C} - mq_e V_s(t) \quad , \tag{10.129}$$

yields the second order discrete equation

$$-\frac{\hbar^2}{2Lq_e^2}(C_{m+1} + C_{m-1}) + \left(\frac{q_e^2}{2C}m^2 - q_e m V_s(t) + \frac{\hbar^2 c^2}{Lq_e^2}\right) C_m(t) = i\hbar\frac{\partial}{\partial t}C_m(t) \quad , \tag{10.130}$$

which generalizes Eq. (6.52) discussed before in connection with the dynamic localization towards the incorporation of an additional harmonic oscillator.

Next let us assume that the voltage is produced in accord with Faraday's law

$$V_s(t) = -\frac{d}{cdt}\Phi_e(t) = \frac{\hbar}{q_e}\mathcal{E}_F f(t) \quad , \tag{10.131}$$

which proceeds in accord with (6.52), where $\Phi_e(t)$ denotes an external time-dependent magnetic flux. The (10.130) produces the Mathieu-type equation

$$\left[-\frac{q_e^2}{2C}\frac{\partial^2}{\partial k^2} + \frac{\hbar^2}{q_e^2 L}\left(1 - \cos\left(k + \frac{q_e\Phi_e}{\hbar c}\right)\right)\right] u(k,t) = i\hbar\frac{\partial}{\partial t}u(k,t) \quad , \tag{10.132}$$

working within the k-representation, where now

$$u(k,t) = \sum_{n=-\infty}^{\infty} C_n(t) \exp\left[in\left(k + \frac{q_e\Phi_e}{\hbar c}\right)\right] \quad , \tag{10.133}$$

and

$$C_n(t) = \frac{1}{2\pi}\sum_{k=-\pi}^{\pi} \exp(-ink) u\left(k - \frac{q_e\Phi_e}{\hbar c}\right) \quad . \tag{10.134}$$

So we found that the effective Hamiltonian characterizing a quantum LC-circuit with a voltage source is given by (Chen et al (2002))

$$\mathcal{H}_{eff}(\Phi_e, k) = -\frac{q_e^2}{2C}\frac{\partial^2}{\partial k^2} + \frac{\hbar^2}{q_e^2 L}\left(1 - \cos\left(k + \frac{q_e\Phi_e}{\hbar c}\right)\right) \quad , \tag{10.135}$$

in accord with (10.132), which proceeds within the k-representation introduced above.

10.6 Dynamic localization effects in L-ring circuits

At this point let us assume that the influence of the capacitance in (10.130) is negligible. The factorization

$$C_m(t) = c_m(t) \exp\left(-i\frac{\hbar c^2}{Lq_e^2}t\right) \tag{10.136}$$

can then be performed, which amounts to remove the related constant potential energy term in (10.130). Next, let us specify the time-dependence of the magnetic flux as follows

$$\Phi_e(t) = \widetilde{\Phi}_0 \sin(\omega t) \quad . \tag{10.137}$$

Then the discrete equation characterizing the rescaled amplitude $c_m(t)$ reproduces identically (6.52) if $\hbar V = -\hbar^2/2Lq_e^2$, $f(t) = \cos(\omega t)$ and $\mathcal{E}_F = -q_e\widetilde{\Phi}_0\omega/c\hbar$. Under such conditions the electron ($q_e = e$) described by the rescaled version of (10.130) is able to exhibit the dynamic localization effect in terms of selected parameters for which

$$J_0\left(\frac{q_e}{c\hbar}\widetilde{\Phi}_0\right) = 0 \quad , \tag{10.138}$$

such as provided by (6.53). Next, we have to realize that

$$I_k = I_k(t) = -c\frac{\partial \mathcal{H}_{eff}}{\partial \Phi_e} = -\frac{\hbar}{q_e L}\sin\left(k + \frac{q_e \Phi_e}{\hbar c}\right) \tag{10.139}$$

has the meaning of a persistent current in the ring. Conversely, (10.139) amounts to consider both a time-dependent and wavenumber realization of the energy. This latter energy can also be obtained by applying the minimal substitution to the energy dispersion law characterizing the L-ring Hamiltonian. It has also been found that the average of he current I_k vanishes in terms of (10.138), which provides a nontrivial manifestation of the dynamic localization. To this aim the MSD has been established in terms of velocity autocorrelation functions (Kenkre et al (1981)).

The time average of the persistent current over one period, say $<I_k(t)>_T$, is of a special interest. This amounts to establish the integral

$$J_k(T) = \int_0^T \sin\left(k + \frac{q_e\Phi_e}{\hbar c}\right) dt \quad , \tag{10.140}$$

in which case $< I_k(t) >_T = -\hbar J_k(T)/q_e TL$. Using again (10.137), it can be easily verified that the DLC (10.138) is reproduced irrespective of k via $J_k(T) = 0$. This means in turn that one should have

$$Z_1(T) = \int_0^T \exp(-i\mathcal{E}_F \eta(t))\, dt = 0 \tag{10.141}$$

in accord with (6.57) and (6.93). The matching condition

$$\Phi_e(t) = \Phi_e(0) - \frac{\mathcal{E}_F \hbar c}{q_e} \eta(t) \tag{10.142}$$

relying on (10.131) has also been considered. In general, $\Phi_e(t)$ in (10.142) may be periodic or not, even if the voltage, or equivalently the $f(t)$-modulation, does it. The interesting point is that (10.141) reproduces identically the exact DLC's characterizing other periodic modulations concerning $f(t)$, like (6.134) and (6.181). Accordingly, (10.141) can be viewed as expressing the exact counterpart of approximate DLC's discussed before in terms of (6.85) and (6.86). In addition, (6.110) and (6.111) show that the extremal condition characterizing the exact MSD gets fulfilled via $D_r(T) = 0$ whenever (10.141) is satisfied. The concavity condition (6.119) remains also valid, provided one inserts again $t = T$ instead of $t = t_g$. Somewhat exceedingly selected modulations for which (10.141) remains valid under all rescalings like $\mathcal{E}_F \to p\mathcal{E}_F$, where p denotes a positive integer, have also been discussed (Dignam and de Sterke (2002); Domachuk et al (2002))). However, such cases are not useful in practice, so that (10.141) stands reasonably for the exact DLC characterizing arbitrary but periodic voltages. In particular, $\eta(t)$ is able to be periodic, too. Then one has $Z_1(nT) = nZ_1(T) = 0$, which results, in general, in a rescaled version of the time grid $t = t_g$ analyzed previously within the strong field limit. However, we can resort from the very beginning to a periodic magnetic flux, in which case the time grid $t_n = nT$ just referred to above gets implemented without extra conditions.

10.7 Double quantum dot systems attached to leads

Double quantum dot systems (DQD's) attached to leads provide further nanodevices which are able to produce non-linear current-voltage characteristic curves working in the Kondo regime at zero temperature (Orellana et al (2002)). The inter-dot Coulomb interaction is neglected, which leads to appreciable simplifications. In addition, a quickly tractable mean field approximation has been applied. This results again in coupled equations for the probability amplitudes, which are rather similar to the ones obtained in the case of other junctions discussed in this chapter.

The dots are localized at the sites $\alpha = 0$ and $\alpha = 1$ of the 1D lattice, while the electron on the left (right) lead is located at the sites $i = -1, -2. -3, ...$ $(i = 2, 3, 4, ...)$. Resorting to the slave boson description (Coleman (1984)) means that the annihilation operator of an electron at dot α is given by

$$\widetilde{c}_{\alpha,\sigma} = b_\alpha^+ f_{\alpha\sigma} = f_{\alpha\sigma} b_\alpha^+ \quad , \tag{10.143}$$

where the boson operator b_α^+ creates an empty state. In this context $f_{\alpha\sigma}$ denotes a fermion operator which is responsible for the annihilation of a single occupied state for which the spin projection is σ in units of $\hbar/2$. The canonical commutation relations read $[b_\alpha, b_\beta] = \delta_{\alpha,\beta}$ and $\{f_{\alpha\sigma}, f_{\beta\mu}\} = \delta_{\alpha\beta}\delta_{\sigma\mu}$. In general, the fermion spin may be different from 1/2. The double occupancy at site α is ruled out by virtue of the constraint

$$Q_\alpha = \sum_\sigma f_{\alpha\sigma}^+ f_{\alpha\sigma} + b_\alpha^+ b_\alpha = 1 \quad . \tag{10.144}$$

This corresponds to the selection of two admissible states like

$$|\alpha, 0 >= b_\alpha^+ |0> \quad , \tag{10.145}$$

and

$$|\alpha, 1 >= f_{\alpha\sigma}^+ |0> \quad , \tag{10.146}$$

such that

$$b_\alpha |0 >= f_{\alpha\sigma} |0 >= 0 \quad . \tag{10.147}$$

It can be easily verified that

$$\widetilde{c}_{\alpha,\sigma} |\alpha, 0 >= 0, \quad \text{and} \quad \widetilde{c}_{\alpha,\sigma}^+ |\alpha, 0 >= |\alpha, 1 > \quad , \tag{10.148}$$

while

$$\widetilde{c}_{\alpha,\sigma}|\alpha,1>=|\alpha,0>, \quad \widetilde{c}^{+}_{\alpha,\sigma}|\alpha,1>=0 \quad . \tag{10.149}$$

So, there is

$$Q_\alpha|\alpha,\varepsilon>=|\alpha,\varepsilon> \quad , \tag{10.150}$$

where $\alpha = 0, 1$, which reflects safely the constraint written down above.

Accordingly, the DQD system is described by the Hamiltonian

$$\mathcal{H}_{QD} = \sum_\alpha \sum_\sigma \varepsilon_\alpha f^{+}_{\alpha\sigma} f_{\alpha\sigma} + \frac{t_c}{2} \sum_\sigma (\widetilde{c}^{+}_{0,\sigma}\widetilde{c}_{1,\sigma} + \widetilde{c}^{+}_{1,\sigma}\widetilde{c}_{0,\sigma}) \quad , \tag{10.151}$$

where t_c and ε_α denote the inter-dot tunneling coupling and the site energy at the dot, respectively. The leads are described by the tight binding Hamiltonian

$$\mathcal{H}_L = \sum_\sigma \left(\sum_{i=-\infty}^{-1} + \sum_{i=2}^{\infty} \right) \varepsilon_i c^{+}_{i,\sigma} c_{i,\sigma} + t \sum_\sigma \left(\sum_{i=-\infty}^{-1} + \sum_{i=3}^{\infty} \right) (c^{+}_{i,\sigma} c_{i-1,\sigma} + c^{+}_{i-1,\sigma} c_{i,\sigma}) \quad , \tag{10.152}$$

with the hopping parameter t, where $c^{+}_{i,\sigma}$ creates an electron with the spin projection $\sigma = \pm 1$ in the site i. The site energies are denoted by ε_α and ε_i, respectively. Next let us denote by V_L (V_R) the hopping between the sites -1 and 0 (1 and 2). Then the Hamiltonian

$$\mathcal{H}_{LQD} = \frac{V_L}{\sqrt{2}} \sum_\sigma (c^{+}_{-1,\sigma}\widetilde{c}_{0,\sigma} + \widetilde{c}^{+}_{0,\sigma} c_{-1,\sigma}) + \frac{V_R}{\sqrt{2}} \sum_\sigma (\widetilde{c}^{+}_{1,\sigma} c_{2,\sigma} + c^{+}_{2,\sigma}\widetilde{c}_{1,\sigma}) \quad , \tag{10.153}$$

describes the hopping interaction between the leads and the QD's. The last step in the construction of the total Hamiltonian is to account for the constraint (10.144) in terms of the additional interaction

$$\mathcal{H}_C = \sum_\alpha \lambda_\alpha (Q_\alpha - 1) \quad , \tag{10.154}$$

where λ_α denotes a Lagrange multiplier. The total Hamiltonian is then given by

$$\mathcal{H}_{tot} = \mathcal{H}_{QD} + \mathcal{H}_L + \mathcal{H}_{LQD} + \mathcal{H}_C \quad , \tag{10.155}$$

which leads, as usual, to the eigenvalue equation

$$\mathcal{H}_{tot}|\Psi >= E|\Psi > \quad . \tag{10.156}$$

Next one applies a mean-field approximation, which amounts to replace the boson operators b_α and b_α^+ by expectation values like

$$< b_\alpha >=< b_\alpha^+ >= \widetilde{b}_\alpha \sqrt{2} \quad . \tag{10.157}$$

This yields the rescaled parameters $\widetilde{\varepsilon}_\alpha = \varepsilon_\alpha + \lambda_\alpha$, $\widetilde{V}_0 = \widetilde{b}_0 V_L$, $\widetilde{V}_1 = \widetilde{b}_1 V_R$ and $\widetilde{t}_c = \widetilde{b}_0 \widetilde{b}_1 t_c$, such that

$$\mathcal{H}_{QD} = \widetilde{\mathcal{H}}_{QD} = \sum_\alpha \sum_\sigma \widetilde{\varepsilon}_\alpha f_{\alpha\sigma}^+ f_{\alpha\sigma} + \widetilde{t}_c \sum_\sigma (f_{0,\sigma}^+ f_{1,\sigma} + f_{1,\sigma}^+ f_{0,\sigma}) \quad , \tag{10.158}$$

and

$$\mathcal{H}_{LQD} = \widetilde{\mathcal{H}}_{LQD} = \widetilde{V}_0 \sum_\sigma (c_{-1,\sigma}^+ f_{0,\sigma} + f_{0,\sigma}^+ c_{-1,\sigma}) + \widetilde{V}_1 \sum_\sigma (f_{1,\sigma}^+ c_{2,\sigma} + c_{2,\sigma}^+ f_{1,\sigma}) \quad . \tag{10.159}$$

This means that the total Hamiltonian exhibits the form

$$\mathcal{H}_{tot} = \widetilde{\mathcal{H}}_{QD} + \mathcal{H}_L + \widetilde{\mathcal{H}}_{LQD} + \sum_\alpha \lambda_\alpha (2\widetilde{b}_\alpha^+ \widetilde{b}_\alpha - 1) \quad , \tag{10.160}$$

which depends on four parameters λ_0, λ_1 ,$\widetilde{b}_0$ and $\widetilde{b}_1$ which will be referred to later in terms of the subscript $j(j = 1 - 4)$.

Proceeding in a close analogy with the previous calculations and choosing again a wavefunction expansion over Wannier states

$$|\Psi >= |\psi_\sigma >= \sum_{l=-\infty}^{\infty} a_{l,\sigma}|l, \sigma > \quad , \tag{10.161}$$

leads to the coupled difference equations

$$(E - \varepsilon_l)a_{l,\sigma} = t(a_{l-1,\sigma} + a_{l+1,\sigma}) \quad , \tag{10.162}$$

$$(E - \varepsilon_l)a_{l,\sigma} = \widetilde{V}_{l+1(l-1)} a_{l+1(l-1),\sigma} + t a_{l-1(l+1),\sigma} \quad , \tag{10.163}$$

and

$$(E-\widetilde{\varepsilon}_l)a_{l,\sigma}=\widetilde{V}_{l-1(l+1)}a_{l+1(l-1),\sigma}+\widetilde{t}_c a_{l+1(l-1),\sigma} \quad , \tag{10.164}$$

for $l\neq -1,0,1,2$, $l=-1,2$ and $l=0,1$, respectively. It is understood that $a_{l+1(l-1),\sigma}=a_{0(1),\sigma}$ if $l=-1,2$ and similarly in other cases.

The four parameters referred to above can be established with the help of the Hellmann-Feynman theorem

$$\partial_j E=<\psi_\sigma|\partial_j\mathcal{H}_{tot}|\psi_\sigma> \quad , \tag{10.165}$$

where ∂_j $(j=1,2,3,4)$ denotes the parameter differentiation needed. Imposing energy minima via $\partial_j E=0$ gives nonlinear parameter-fixing conditions like

$$2\widetilde{b}_\alpha^2+\sum_\sigma |a_{\alpha,\sigma}|^2=1 \quad , \tag{10.166}$$

and

$$2\lambda_\alpha\widetilde{b}_\alpha^2+\widetilde{V}_\alpha\sum_\sigma Re(a^*_{\alpha-1(\alpha+1),\sigma}a_{\alpha,\sigma})+\widetilde{t}_c\sum_\sigma Re(a^*_{\alpha+1(\alpha-1),\sigma}a_{\alpha,\sigma})=0 \quad , \tag{10.167}$$

where $\alpha=0,1$. Further details can then be established by discussing (10.162)-(10.164) in terms of plane wave realizations of the probability amplitudes, as done in the case of other junctions presented before. Just mention that handling numerically the above equations leads to nontrivial interplays between the bistability of the current and the Kondo-effect (Orellana et al (2002)).

Chapter 11

Conclusions

In this volume we learned how to deal with the theoretical description of the Schrödinger-equation on the discrete space, with a special emphasis on applications to low-dimensional nanoscale structures. Almost cases concern solvable systems on the discrete space, but time discretizations have also been discussed. Other useful generalizations, like matrix versions of the discrete Schrödinger operator, have to be mentioned (Bruschi et al (1981)). Underlying symmetries play an important role and serve both for a better understanding as well as for the formulation of suitable calculation techniques. First we would like to address some mathematical developments. We emphasize that such developments are able to serve as candidates for further applications. So representations of the q-Heisenberg algebra (Pan and Zhao (2001), Curado et al (2001)) are able to be applied to the derivation of explicit eigenvalues and spectral properties. This algebra plays actually an essential role in the description of discretized systems (Celeghini et al (1995)), while noncanonical Heisenberg algebras look promising towards a better understanding of q-deformations (Brodimas et al (1992), Janussis and Brodimas (2000)). The harmonic oscillator has been generalized to the relativistic configuration space proposed by Kadyshevsky et al (1968). We have to recall that eigenvalue equations on this space look like non-relativistic difference Schrödinger-ones, but now with a relativistic coordinate, as indicated in section 4.6. This leads to. a q-deformed oscillator, for which the dynamical symmetry is given by the quantum group $SU_q(1,1)$. The corresponding deformation parameter is given by

$$q = \exp\left(- \frac{\hbar\omega_0}{4m_0c^2} \right) \tag{11.1}$$

where ω_0 denotes the oscillator frequency. Correspondingly, the annihilation and creation operators are themselves finite-difference ones (Mir-

Kasimov (1991)). For such problems, the appropriate finite-difference generalization of the factorization method has already been done by Kagramanov et al (1990).

The main body of this book concerns, however, hopping Hamiltonians and tight binding descriptions under the influence of external fields. The typical example is the celebrated Harper-equation which serves to the description of the $2DEG$ in the magnetic field, i.e. of Bloch electrons on a $2D$ lattice threaded by a transversal and homogeneous magnetic field. The q-symmetrized version of the Harper-equation, which is also interesting from a theoretical point of view, has been analyzed by accounting in an explicit manner for the appropriate Coulomb gauge. The incorporation of an additional electric field into the Harper-equation (8.9) is of a special interest (see also Muñoz et al (2005)). Choosing a time dependent electric field directed along the Ox-axis and using the vector potential

$$A_1(t) = -\hbar c \mathcal{E}_F \eta\,(t)\,/ea \tag{11.2}$$

gives the time dependent magneto-electric Harper-equation

$$\exp\left(i\theta_F\right)u_{n+1} + \exp\left(-i\theta_F\right)u_{n-1} + 2\Delta\cos\left(2\pi\beta n + \theta_2\right)u_n = i\frac{\partial}{\partial t}u_n \tag{11.3}$$

in accord with (6.57), (8.1) and (8.4), where

$$\theta_F = \theta_1 - \mathcal{E}_F \eta\,(t) \quad . \tag{11.4}$$

It is understood that the dimensionless time variable used above is measured in units of $\hbar/\mathcal{E}_0$. The time dependent equation established in this manner serves to the study of carrier propagation under electric and magnetic fields. However, the systematic derivation of exact analytic solutions to (11.3) is still open for further investigations. Just remark that the Hamiltonian characterizing (11.3) is generated via (8.9) by inserting θ_F instead of θ_1. Other low dimensional systems on the discrete space, like the $1D$ conductor in the presence of electric fields, discretized and continuous $1D$ rings threaded by a magnetic flux, quantum LC-circuits and especially junctions between rings and leads or rings and dots, have received much attention. Such systems look promising as they are able to serve as prototypes in the design of nanodevices.

Total persistent currents in the discretized Aharonov-Bohm rings have been derived in an explicit manner. Resorting to Fourier series, we succeeded to establish generalized current formulae exhibiting interesting odd-even parity effects in terms of the number of electrons. We found it useful to present a thorough study of the dynamic localization of charged particles on the 1D lattice under the influence of time dependent electric fields. The possibility to establish leading approximations to DLC's for arbitrary periodic modulations in a rather reasonable manner has been discussed. Dynamical localization effects have also been found in terms of zero average values of ring currents (Chen et al (2002)). Other quasiperiodic driven systems, like kicked rotor Hamiltonians (see e.g. Moore et al (1994)), lead specifically to dynamic localization effects, too. Such systems have been realized in the laboratory by using cold cesium atoms in a pulsed standing wave of light (Klappauf et al (1998), Ringot et al (2000)). In this latter context, the dynamical localization can be understood as a quantum suppression of diffusion in systems that are classically chaotic. The kicked version of the Harper-equation has also been discussed in several respects (Lévi and Georgeot (2004)). The influence of time dependent external fields deserves further attention, which concerns both transport properties and dynamic localization effects. Couplings to quantized microwave fields have also to be accounted for (Migliore and Messina (2004)). Useful steps have also been done in the nonlinear description of transport phenomena (Wacker (2002)).

The exact DOS's established for the Harper-equation in section 8.4 opens the way to a systematic study of thermodynamic properties in the presence of the anisotropy parameter. However, we have to realize that handling related elliptic integrals is not an easy matter. Transport properties concerning the conductance, have been addressed in a close connection with nanodevice junctions mentioned above. Useful details concerning the quantum Hall effect, or the localization length, have also been considered. Besides accounting for disorder and interaction effects (Németh and Pichard (2005), Carvalho Dias et al (2006)), the incorporation of dissipation effects (Guinea (2002)) deserves further attention, too. In addition, the influence of the spin-orbit interaction (Chen et al (2006); Zhang and Xia (2006)), the increasing role of optical lattices (Jaksch and Zoller (2003), Ruuska and Törmä (2004)) and especially the advent of quantum computing (Nielsen and Chuang (2000)), has attracted a lot of interest during the last years. The sensible point is that the spin-orbit interaction is always present in low dimensional structures, like quantum dots and quantum wires (Stormer et al (1983), Bellucci and Onorato (2005)). In the case of 2D electrons, this latter interaction relies on the Rashba-effect (see e.g. Bellucci and Onorato (2003)). Such developments open the way to a new field of research dealing with the implementation of the so called spin Hall effect (Hirsch (1999), Bellucci and Onorato (2003), Raimondi and Schwab (2005)). Moreover, it has been realized that spin dependent transport phenomena in nanostructures are of increasing interest in semiconductor electronics (Awschalom et al (2002)). Other developments in the field of mesoscopic systems have been reviewed recently (Akkermans and Montambaux (2004)), with a special emphasis on quantum diffusion effects.

11.1 Further perspectives

Besides spin chains (Gaudin (1983), Korepin et al (1992)) a rather special class of discretized models is represented by cellular automata, for which appreciable progress has been done during the last two decades (Wolfram (1986), Chapard and Drozi (1998)). Accounting for electron-electron interactions results in Hubbard-models, i.e. in hopping Hamiltonians supplemented by non-linear four-Fermi terms (Gebhard (1997)). The discretized Boltzmann-equation has been applied in the description of microfluids (Li and Kwok (2003)), but the lattice dynamics of strongly correlated systems is also worthy of being mentioned (Savrasov and Kotliar (2003)). Reaction diffusion systems, with NN-interactions on a $1D$ lattice have been solved exactly by applying the generalized empty internal method (Aghamohammadi et al (2003)). Going beyond non-relativistic quantum mechanics there is a broad research field ranging from discretized Dirac (Chakraborti (1995), Lorente and Kramer (1999), Hayashi et al (1997)) and Klein-Gordon (Negro and Nieto (1996)) equations, to discrete sine-Gordon (Korepin et al (1992), Watanabe et al (1995)) and discrete nonlinear Schrödinger equations (Ablowitz and Clarkson (1991), Hoffmann (2000)). Discrete counterparts of many other nonlinear systems and, in particular, of the Korteweg-de Vries equation (Levi and Rodriguez (1999)), have also been discussed. Having established discrete formulations of the Dirac and/or Klein-Gordon equations serves as a starting point to the derivation of relativistic manifestations of dynamic localization effects (see, for instance, de Oliveira et al (2005)). Moreover, the quantum Hall effect has been studied recently in terms of the (2+1)-dimensional massless Dirac equation (Gusynin and Sharapov (2005), Beneventano and Santangelo (2006)).

Noncommutative spaces characterized by commutation relations of the type (Connes (1994) Seiberg and Witten (1999))

$$[X_j, X_k] = i\Theta_{jk} \tag{11.5}$$

which differs from (1.56), have also their own interest. Now the noncommutative parameter is denoted by Θ_{jk}, such that $\Theta_i = \varepsilon_{ijk}\Theta_{jk}$. It has been found that (11.5) produces corrections to the Lamb-shift (Chaichian et al (2001)), but other cases are of a special interest, too. So (11.5) induces an additional momentum-dependent potential such as given by

$$V_{int}^{(\theta)} = \frac{e}{4\hbar}\overrightarrow{\Theta} \cdot (\overrightarrow{p}_{op} \times \overrightarrow{E}) \tag{11.6}$$

under the influence of an electric field, as shown recently in the study of the noncommutative Stark-effect (Chair et al (2005)). The influence of a transversal magnetic field on electrons moving on a $2D$ noncommutative plane has also been discussed (Jellal (2001) and references therein). Besides being applied again to the quantum Hall effect (Bellissard et al (1994)), the noncommutative space is able to produce specifically sensible corrections and/or refinements, modified systems on the discrete space included.

Appendix A

Dealing with polynomials of a discrete variable

In this Appendix we would like to present some additional relationships for readers interested in discrete polynomials. We then have to realize that a polynomial solution $y(x) = y_n(x)$ to (1.67) leads to the eigenvalue

$$\lambda = \lambda_n = -n\left(\tau\prime(x) + \frac{n-1}{2}\sigma\prime\prime(x)\right) \tag{A.1}$$

in which case μ_m becomes

$$\mu_m \to \mu_{mn} = (m-n)\left(\tau\prime(x) + \frac{n+m-1}{2}\sigma\prime\prime(x)\right) \tag{A.2}$$

for $m = 0, 1, 2, ..., n-1$. Proceeding further, the Rodriguez formula

$$y_n(x) = \frac{B_n}{\rho(x)}\nabla^n(\rho_n(x)) \tag{A.3}$$

in which

$$y_n(x) = a_n x^n + b_n x^{n-1} + ... \quad . \tag{A.4}$$

plays a crucial role. There is

$$B_n = \frac{\Delta^n y_n(x)}{A_{nn}} \tag{A.5}$$

where

$$\Delta^n y_n(x) = y_n^{(n)}(x) = n! a_n \tag{A.6}$$

and

$$A_{nn} = n! \prod_{l=0}^{n-1} \left(\tau\prime(x) + \frac{n+l-1}{2}\sigma\prime\prime(x) \right) \quad . \tag{A.7}$$

In order to establish the coefficient a_n we have to resort, however, to a fixing condition concerning B_n. Accordingly, $a_n = ((\mu - 1)/\mu)^n$ and $a_n = 1/n!$ for Meixner- and Krawtchouk polynomials, respectively. Having obtained a_n yields b_n as

$$b_n = \frac{\tau(0) + (n-1)(\sigma\prime(0) + \tau\prime(x)/2)}{\tau\prime(x) + (n-1)\sigma\prime\prime(x)} n a_n \quad . \tag{A.8}$$

Other points which remain to be clarified are the normalization and recursion relations. For finite b-values the $y_n(x)$ polynomial is normalized as

$$\sum_{x=a}^{b-1} y_m(x) y_n(x) \rho(x) = d_n^2 \delta_{m,n} \tag{A.9}$$

where $x = x_i$, which works in combination with the boundary conditions

$$\sigma(x)\rho(x)x^l \mid_{x=a,b} \tag{A.10}$$

and

$$\rho(x) > 0 \tag{A.11}$$

where $a \leq x = x_i \leq b - 1$. In practice, one takes $a = 0$, so that $\sigma(0) = 0$. The normalization constant is then given by

$$d_n^2 = (-1)^n A_{nn} B_n^2 \rho_{b-1}(0) \prod_{k=n}^{b-2} \frac{1 + \sigma\prime\prime/2\tau_k\prime}{\sigma(x_k^*)} \tag{A.12}$$

in which x_k^* denotes the root of $\tau_k(x) = 0$. For Krawtchouk polynomials there is $b \equiv N_0 + 1$, but $b \to \infty$ for Meixner ones. Then the normalization constants are given by

$$d_n^2 = \frac{N_0!}{n!(N_0 - n)!}(pq)^n, \quad \rho(x) = \frac{N_0! p^x q^{N_0 - x}}{\Gamma(x+1)\Gamma(N_0 + 1 - x)} \tag{A.13}$$

and

$$d_n^2 = \frac{n!\Gamma(x+\gamma)}{\mu^n(1-\mu)^\gamma\Gamma(\gamma)}, \quad \rho(x) = \frac{\mu^x\Gamma(\gamma+x)}{\Gamma(x+1)\Gamma(\gamma)} \tag{A.14}$$

respectively. The corresponding weight functions have also been inserted.

The recurrence relation is given by

$$xy_n(x) = \alpha_n y_{n+1}(x) + \beta_n y_n(x) + \gamma_n y_{n-1}(x) \tag{A.15}$$

where

$$\gamma_n = \frac{a_{n+1}}{a_n}\frac{d_n^2}{d_{n-1}^2}, \quad \alpha_n = \frac{a_n}{a_{n+1}} \tag{A.16}$$

and

$$\beta_n = \frac{b_n}{a_n} - \frac{b_{n+1}}{a_{n+1}} \quad . \tag{A.17}$$

A special interest concerns the relationship

$$\left[\sigma(x)\nabla - \frac{\lambda_n}{n\tau_n'}\tau_n(x)\right] y_n(x) = -\frac{\lambda_n}{n\tau_n'}\frac{B_n}{B_{n+1}} y_{n+1}(x) \tag{A.18}$$

which serves to the definition of a related raising operator. A lowering operator is then easily derived by inserting instead of y_{n+1} in (A.18) the equivalent expression indicated by (A.15). In the case of Krawtchouk polynomials one has $\alpha_n = n+1$, $\beta_n = n + p(N_0 - 2n)$ and $\gamma_n = pq(N_0 - n + 1)$, whereas

$$b_n = -\frac{1}{(n-1)!}\left[N_0 p + \frac{1}{2}(n-1)(q-p)\right] \quad . \tag{A.19}$$

The discrete equation and the recurrence relation characterizing normalized Krawchouk-functions (see also (5.41)) are given by

$$\begin{aligned} &\sqrt{pq(x+1)(N_0-x)}K_n^{(p)}(x+1,N_0) + \sqrt{pqx(N_0-x+1)}K_n^{(p)}(x-1,N_0)+ \\ &\quad +[n - N_0 p + x(p-q)]K_n^{(p)}(x,N_0) = 0 \end{aligned} \tag{A.20}$$

and

$$[N_0p+(q-p)n-x]K_n^{(p)}(x,N_0)+\sqrt{pq(n+1)(N_0-n)}K_{n+1}^{(p)}(x,N_0)+ \tag{A.21}$$
$$\sqrt{pqn(N_0-n+1)}K_{n-1}^{(p)}(x,N_0)=0$$

respectively. It is then an easy exercise to verify e.g. that the normalized counterparts of (5.55) and (5.57) becomes

$$p(x-N_0+n)K_n^{(p)}(x,N_0)+\sqrt{pqx(N_0-x+1)}K_n^{(p)}(x-1,N_0)= \tag{A.22}$$
$$=\sqrt{pq(n+1)(N_0-n)}K_{n+1}^{(p)}(x,N_0)$$

and

$$q(x-n)K_n^{(p)}(x,N_0)-\sqrt{pqx(N_0-x+1)}K_n^{(p)}(x-1,N_0)= \tag{A.23}$$
$$=\sqrt{pqn(N_0-n+1)}K_{n-1}^{(p)}(x,N_0)\quad .$$

This shows that related raising and lowering operators can be introduced in a well defined manner, as indicated by Lorente (2001b).

The recurrence relations characterizing the Meixner-polynomials and the normalized Meixner-function read

$$(x(\mu-1)+n+\mu(n+\gamma))m_n^{(\gamma,\mu)}(x)=\mu m_{n+1}^{(\gamma,\mu)}(x)+n(n-1+\gamma)m_{n-1}^{(\gamma,\mu)}(x) \tag{A.24}$$

and

$$(x(\mu-1)+n+\mu(n+\gamma))M_n^{(\gamma,\mu)}(x)=\sqrt{\mu(n+\gamma)(n+1)}M_{n+1}^{(\gamma,\mu)}(x)+ \tag{A.25}$$
$$+\sqrt{\mu n(n-1+\gamma)}M_{n-1}^{(\gamma,\mu)}(x)$$

respectively. The corresponding discrete equations are given by (5.62) and (5.64). Such results are of interest in the study of factorization properties.

Without belaboring our presentation, we have just to say that the Charlier polynomials $c_n^{(\mu)}(x)$, where $\mu>0$, are defined for $x\in[0,\infty)$, in which case the weight function and the normalization constant are given by

$$\rho(x)=\frac{\mu^x\exp(-\mu)}{\Gamma(x+1)} \tag{A.26}$$

and

$$d_n = \left[\frac{n!}{\mu^n} \exp(-\mu) \right]^{1/2} \tag{A.27}$$

respectively. These polynomials satisfy the discrete equation

$$\mu c_n^{(\mu)}(x+1) + x c_n^{(\mu)}(x-1) + (n - \mu - x) c_n^{(\mu)}(x) = 0 \tag{A.28}$$

whereas the corresponding recurrence relation reads

$$(\mu + n - x) c_n^{(\mu)}(x) = \mu c_{n+1}^{(\mu)}(x) + n c_{n-1}^{(\mu)}(x) \quad . \tag{A.29}$$

It has been shown by Atakishiyev and Suslov (1990), that such polynomials serve to the description of a particular realization of a difference harmonic oscillator. This originates from the discrete equation characterizing the normalized Krawtchouk functions via $N_0 \to \infty$ and $p \to \mu/N$, where μ stands for a fixed parameter. The normalized counterparts of such polynomials are given by

$$C_n^{(\mu)}(x) = \sqrt{\frac{\mu^n \exp(-\mu) \mu^x}{n! x!}} c_n^{(\mu)}(x) \quad . \tag{A.30}$$

The normalized polynomial established in this manner obeys the discrete equation

$$\sqrt{\mu(x+1)} C_n^{(\mu)}(x+1) + \sqrt{\mu x} C_n^{(\mu)}(x-1) + (n - x - \mu) C_n^{(\mu)}(x) = 0 \tag{A.31}$$

whereas the recurrence relation reads

$$\sqrt{\mu(n+1)} C_{n+1}^{(\mu)}(x) + \sqrt{\mu n} C_{n-1}^{(\mu)}(x) + (x - n - \mu) C_n^{(\mu)}(x) = 0 \quad . \tag{A.32}$$

A further point of interest is the connection of orthogonal polynomials with hypergeometric functions. So the Krawtchouk, Meixner and Charlier polynomials exhibit the realizations (Nikiforov et al (1991))

$$k_n^{(p)}(x, N_0) = \frac{(-p)^n}{n!} \frac{\Gamma(N_0 - x + 1)}{\Gamma(N_0 - x - n + 1)} {}_2F_1(-n, -x, N_0 - x - n + 1; -q/p) \tag{A.33}$$

$$m_n^{(\gamma,\mu)}(x) = \frac{\Gamma(n + \gamma)}{\Gamma(\gamma)} {}_2F_1(-n, -x, \gamma; 1 - 1/\mu) \tag{A.34}$$

and

$$c_n^{(\mu)}(x) = \frac{(-1)^n}{\mu^n} \frac{\Gamma(x+1)}{\Gamma(x-n+1)} \, {}_1F_1(-n, x-n+1; \mu) \tag{A.35}$$

respectively. Such relationships are useful for the derivation of duality properties like

$$k_n^{(p)}(m, N_0) = (-p)^{n-m} \frac{m!(N_0-m)!}{n!(N_0-n)!} \, k_n^{(p)}(n, N_0) \tag{A.36}$$

$$m_n^{(\gamma,\mu)}(l) = \frac{\Gamma(\gamma+n)}{\Gamma(\gamma+l)} \, m_l^{(\gamma,\mu)}(n) \tag{A.37}$$

and

$$c_n^{(\mu)}(m) = c_m^{(\mu)}(n) \tag{A.38}$$

which work for $n, l, m = 0, 1, 2, ...$

Further results can be derived by combining (A.33)-(A.35) with Gauss's relations for contiguous hypergeometric functions such as given by (see e.g. (15.2.10) in Abramowitz and Stegun (1972))

$$\begin{aligned}(c-a){}_2F_1(a-1,b,c;z) + (2a-c-az+bz)\,{}_2F_1(a,b,c;z) + \\ +a(z-1)\,{}_2F_1(a+1,b,c;z) = 0 \quad .\end{aligned} \tag{A.39}$$

Such hypergeometric functions satisfy the differential equation

$$z(z-1)\frac{d^2u}{dz^2} + (c-(a+b+1)z)\frac{du}{dz} - abu = 0 \tag{A.40}$$

which relies on (1.39) via $u = F(a, b, c : z)$. Generalized orthogonal polynomials are also able to be provided, in a rather different manner, by applying the Jacobi-matrix method in conjunction with square integrable Sturm-Laguerre basis functions (Yamani and Fishman (1975)). It should be mentioned that the Jacobi-matrix is tridiagonal, which leads to "sieved" orthogonal polynomials referred to above in terms of corresponding three-term recurrence relations. Such basis functions can also be converted to the ones characterizing the isotropic harmonic oscillator by virtue of Coulomb-oscillator duality (see e.g. Kostelecky and Russell (1996)). Under such circumstances the three-dimensional radial Schrödinger-equations with Coulomb- and harmonic oscillator potentials leads to Pollaczek- and Meixner-polynomials, respectively, as shown in Bank and Ismail (1985). A relativistic generalization of this method has also been proposed recently (Alhaidari (2002)).

Appendix B

The functional Bethe-ansatz solution

Another special problem for the reader interested in theoretical developments is the derivation of an implicit solution to the q-difference equation (Wiegmann and Zabrodin (1994a), Wiegmann and Zabrodin (1995))

$$\mathcal{H}\psi(z) = a(z)\,\psi\left(q^2 z\right) + d(z)\,\psi\left(\frac{z}{q^2}\right) - v(z)\,\psi(z) = E\psi(z) \tag{B.1}$$

has been done by applying an algebraization approach which is reminiscent to the so called Bethe-ansatz method serving to the description of integrable models in statistical physics and field theory (Korepin et al (1992)) . For this purpose a suitable class of Hamiltonians has been considered by choosing the coefficient functions as

$$a(z) = \frac{b_1}{q^{4j-1}} z^2 - \frac{b_3}{q^{3j}} z + \frac{a}{q^{2j}} + \frac{c_2}{q^j z} + \frac{c_1}{q z^2} \tag{B.2}$$

$$d(z) = b_1 q^{4j-1} z^2 + b_2 q^{3j} z + d q^{2j} - \frac{c_3 q^j}{z} + \frac{c_1 q}{z^2} \tag{B.3}$$

and

$$v(z) = \left(q + \frac{1}{q}\right) b_1 z^2 + \left(\frac{b_2}{q^j} - b_3 q^j\right) z + \tag{B.4}$$

$$+ \left(\frac{c_2}{q^j} - c_3 q^j\right) \frac{1}{z} + \frac{c_1}{z^2}\left(q + \frac{1}{q}\right) \quad .$$

It is understood that at least one the coefficients c_1, c_2 and c_3 is nonzero.

Now one starts from a polynomial wavefunction like

$$\psi(z) = \prod_{m=1}^{\widetilde{N}} (z - z_m) \tag{B.5}$$

in which the z_m-roots remain to be determined by ruling out singularities. implied in the quotient

$$E(z) = \frac{1}{\psi(z)} \mathcal{H}\psi(z) \quad . \tag{B.6}$$

Indeed, inserting (B.5) into (B.1) gives

$$E(z) = a(z) \prod_{m=1}^{\widetilde{N}} \frac{q^2 z - z_m}{z - z_m} + d(z) \prod_{m=1}^{\widetilde{N}} \frac{z/q^2 - z_m}{z - z_m} - v(z) = E \tag{B.7}$$

so that there are double and simple poles at $z = 0$, $z = \infty$, as well as the z_m-poles implied by the denominators characterizing $E(z)$. The singular contributions at $z = \infty$ are incorporated into the $E(z)$ -terms proportional to z^2 and z:

$$E_S(z) = b_1 \left(q^{2\widetilde{N}-4j+1} + \frac{1}{q^{2\widetilde{N}-4j+1}} - q - \frac{1}{q} \right) z^2 + \tag{B.8}$$

$$+ b_2 \left(q^{-2\widetilde{N}+3j} - \frac{1}{q^j} \right) z + b_3 \left(q^j - q^{2\widetilde{N}-3j} \right) +$$

$$+ b_1 z \left(q - \frac{1}{q} \right) \left(q^{4\widetilde{N}-4j} - \frac{1}{q^{2\widetilde{N}-4j}} \right) \sum_{m=1}^{\widetilde{N}} z_m \quad .$$

One sees that $E_S(z)$ is ruled out if

$$\widetilde{N} = 2j \tag{B.9}$$

so that the degree of the polynomial is well determined if at least one of the coefficients b_1, b_2 and b_3 is nonzero. It is furthermore clear that

$$E(z) \to \frac{a}{q^{2j}} + dq^{2j} + c_1 \frac{\left(1 - q^2\right)^2 \left(1 + q^2\right)}{q^3} \sum_{l>m}^{\widetilde{N}} \frac{1}{z_m z_l} + \tag{B.10}$$

$$+\left(1-q^2\right)\left(\frac{c_2}{q^j}+c_3q^{j-2}\right)\sum_{m=1}^{\widetilde{N}}\frac{1}{z_m}$$

as $z \rightarrow 0$, which means that the singular part at $z=0$ is ruled out from the very beginning. We have also to remark that

$$E(z) \rightarrow aq^{2(\widetilde{N}-j)}+\frac{d}{q^{2(\widetilde{N}-j)}}+ \tag{B.11}$$

$$+b_1\left(q^{2\widetilde{N}-4j-3}+\frac{1}{q^{2\widetilde{N}-4j+1}}\right)\left(1-q^2\right)^2\sum_{l>m}^{\widetilde{N}}z_lz_m+$$

$$+\left(\frac{b_2}{q^{2\widetilde{N}-3j}}+b_3q^{2\widetilde{N}-3j-2}\right)\left(1-q^2\right)\sum_{m=1}^{\widetilde{N}}z_m$$

as $z \rightarrow \infty$, which proceeds in accord with (B.9) and which sheds some light on the role of coefficients a, d, b_1, b_2 and b_3 in the energy description.

The mutual annihilation of the $z=z_m$-poles is done in terms of the condition

$$q^{-2\widetilde{N}}\frac{d(z_l)}{a(z_l)}=\prod_{\substack{m=1\\ m\neq l}}^{\widetilde{N}}\frac{q^2z_l-z_m}{z_l-q^2z_m} \tag{B.12}$$

where $l=1,2,\ldots,\widetilde{N}$. Putting together the constant terms yields the energy eigenvalue as

$$E=b_1\left(q-\frac{1}{q}\right)\left(q^2-\frac{1}{q^2}\right)\sum_{n<m}^{\widetilde{N}}z_nz_m- \tag{B.13}$$

$$-\left(q-\frac{1}{q}\right)\left(\frac{b_2}{q^{j-1}}+b_3q^{j-1}\right)\sum_{m=1}^{\widetilde{N}}z_m+aq^{2j}+\frac{d}{q^{2j}}$$

which resembles very much (B.11). Accordingly, all that one needs in order to establish the energy and the wavefunction is to solve (B.12), but in practice this is an intricate problem. This technique has been applied in the study of the qSHE (Wiegmann and Zabrodin (1994a), Wiegmann and Zabrodin (1994b), Hatsugai et al (1994)).

We have to remark, that (B.1) can be converted into a second-order discrete equation via $z = q^{2n}$ and $\psi\left(q^{2n}\right) = \psi_n$ and similarly for the coefficient functions. So (B.1) becomes

$$\mathcal{H}\psi_n = a_n\psi_{n+1} + d_n\psi_{n-1} - v_n\psi_n = E\psi_n \tag{B.14}$$

which is able to be related both to (1.10) and (11.1).

Bibliography

Abanov A G, Talstra J C and Wiegmann P B 1998 Asymptotically exact wave functions of the Harper equation *Phys. Rev. Lett.* **81** 2112-2115

Ablowitz M J and Clarkson P A 1991 *Solitons, Nonlinear Evolutions Equations and Inverse Scattering* (Cambridge: Univ. Press.)

Abrahams E, Anderson P W, Licciardello D C and Ramakrishnan T V 1979 Scaling theory of localization: absence of quantum diffusion in two dimensiona *Phys. Rev. Lett.* **42** 673-676

Abramowitz M and Stegun I A 1972 *Handbook of Mathematical Functions*(New York: Dover)

Aghamohammadi A, Alimohammadi M and Khorrami M 2003 Exactly solvable models through the generalized empty interval method *Eur. Phys. J.* B **31** 371-378

Aharonov Y and Bohm D 1959 Significance of elecromagnetic potential in the quantum theory *Phys.Rev* **115** 485-491

Akkermans E, Auerbach A, Avron J E Shapiro B 1991 Relation between persistent currents and the scattering matrix *Phys. Rev. Lett.* **66** 76-79.

Akkermans E and Montambaux G 2004 *Physique Mésoscopique des Electrons et des Photons* (Paris: EDP Sciences).

Alavi S A and Rouhani S 2004 Exact analytical expression for magnetoresistance using quantum groups *Phys. Lett.* A **320** 327-332

Albrecht C, Smet J H, von Klitzing K, Weiss D, Umansky V and Schweizer H 2001 Evidence of Hoftadter's fractal energy spectrum in quantized hall conductance *Phys. Rev. Lett.* **86** 147-150.

Alexandrov A S and Bratkovsky A M 2001 Semiclassical theory of magnetic quantum oscillations in a two-dimensional multiband canonical Fermi-liquid *Phys. Rev.* B **65** 033105 pp. 1-3

Alhaidari A D 2002 The relativistic J-matrix theory of scattering: an analytic solution *J. Math. Phys.* **43** 1129-1135

Ambarzumian V and Iwanenko D 1930 Zur Frage nach der Vermeidung der unendlichen Selbstrückwirkung des Electrons *Z. Phys.* **64** 563-567

Anderson P W 1958 Absence of diffusion in certain random lattices *Phys. Rev.* **109** 1492-1505

Ando T, Fowler A B and Stern F 1982 Electronic properties of two-dimensional systems *Rev. Mod. Phys.* **54** 437-672

Apenko S M 1998 Environment induced decompactification of phase in Josephson junctions *Phys. Lett.* A **142** 277-281

Ashcroft N W and Mermin N D 1976 *Solid State Physics* (Philadelphia: Saunders)

Atakishiyev N M and Suslov S K 1990 Difference analogs of the harmonic oscillator *Theor. Math. Phys.* **85** 1055-1062

Atakishiyev N M and Wolf K N 1997 Fractional Fourier-Krawtchouk transform *J. Opt. Soc. Am. A* **14** 1467-1472

Aubry S and Andre G 1980 Analicity breaking and Anderson localization in incommensurate lattices *Ann. Isr. Phys. Soc.* **3** 133-164

Aunola M 2003 The discretized harmonic oscillator: Mathieu functions and a new class of generalized Hermite polynomials *J. Math. Phys.* **44** 1913-1936

Avgin I 2002 The CPA solution of the anomalous Lyapunov exponent for disordered binary chains *Eur. Phys. J.* B **29** 437-440

Avishai Y, Hatsugai Y and Kohmoto M 1993 Persistent currents and edge states in a magnetic field *Phys. Rev.* B **47** 9501-9512

Avron J E, Mouche van P and Simon B 1990 On the measure of the spectrum for the almost Mathieu operator *Commun. Math. Phys.* **132** 103-118

Awschalom D D, Loss D and Samarth N 2002 *Semiconductor Spintronics and Quantum Computation* (Berlin:Springer)

Azbel M Ya 1979 Quantum particle in one-dimensional potentials with incommensurate periods, *Phys. Rev. Lett.* **43** 1954-1957

Babelon O, Bernard D and Billey E 1996 A quasi-Hopf algebra interpretation of quantum 3-j and 6-j symbols and difference equations *Phys. Lett.* B **375** 89-97

Bagchi B and Ganguly A 2003 A unified treatment of exactly solvable and quasi-exactly solvable quantum potentials *J. Phys.* A **36** L161-L167

Bailey W 1935 *Generalized Hypergeometric Series* (London: Cambridge Univ. Press)

Bank E and Ismail M E N 1985 The attractive Coulomb potential polynomials *Constr. Approx.* **1** 103-119

Barticevic Z, Fuster G and Pacheko M 2002 Effect of an electric field on the Bohn-Aharonov oscillations in the electronic spectrum of a quantum ring *Phys. Rev.* B **65** 193307 pp. 1-4

Bastard G, 1992 *Wave mechanics applied to semi-conductor heterostructures* (Paris: EDP Sciences)

Beck G 1929 Die zeitliche Quantelung der Bewegung *Z. Phys.* **51** 737-739

Bellisard J A, Elst van A and Schulz-Baldes H 1994 The noncommutative geometry of the quantum Hall-efect *J. Math. Phys.* **35** 5373-5451

Belluci S and Onorato P 2003 Rasba effect in two-dimensional mesoscopic systems with transverse magnetic field *Phys. Rev.* B **68** 245322 pp. 1-11

Bellucci S and Onorato P 2005 Spin-orbit coupling in a quantum dot at high magnetic field *Phys. Rev.* B **72** 045345 pp. 1-7

Bender C M, Milton K A Sharp D H, Simmons Jr. L M and Stong R 1985 Discrete time quantum mechanics *Phys. Rev. D* **32** 1476-1485

Bender C M and Dunne G V 1988 Exact operator solutions to Euler Hamiltonians *Phys. Lett.* **200** 520-524

Bentosela F, Briet Ph and Pastur L 2003 On the spectral and wave propagation properties of the surface Maryland model *J. Math. Phys.* **44** 1-35

Beneventano C G and Santangelo E M 2006 Relativistic Landau problem at finite temperature *J. Phys.* A **39** 6137-6144

Besicovitch A S 1932 *Almost Periodic Functions*(London: Cambridge Univ. Press)

Biedenharn L C 1989 The quantum group $SU_q(2)$ and a q-analogue of the boson operators *J. Phys. A* **22** L873-L878

Bleuse J, Bastard G and Voisin P 1988 electric fields induced localization and oscillatory electro-optical properties of semiconductor superlattices *Phys. Rev. Lett* **60** 220-223.

Bloch F 1928 Über die Quantenmechanik der Elektronen in Kristallgittern, *Z. Phys.* **52** 555-600.

Boiti M, Bruschi M, Pempinelli F and Prinari B 2003 A discrete Schrödinger spectral problem and associated evolution equations *J. Phys. A* **36** 139-149

Bonatsos D, Daskaloyannis C, Ellines D and Faessler A 1994 Discretization of the phase space for a q-deformed harmonic oscillator with q a root of unity *Phys. Lett. B* **331** 150-156

Bonifacio R and Caldirola P 1983 Finite difference equation and quasi-diagonal form in quantum statistical mechanics *Lett. Nuovo Cimento* **38** 615-619

Bonifacio R 1983 A coarse grained description of time evolution: irreversible state reduction and time-energy relation *Lett. Nuovo Cimento* **37** 481-489

Boon M H 1972 Representations of the invariance group for a Bloch electron in a magnetic field *J. Math. Phys.* **13** 1268-1284

Boykin T B and Klimeck G 2004 The discretized Schrödinger equation and simple models for semiconductor quantum wells *Eur. J. Phys.* **25** 503-514

Brill D R and Gowdy R H 1970 Quantization of the general theory of relativity *Rep. Progr. Phys.* **33** 413-510

Brinkman W F and Rice T 1970 Single-particle excitations in magnetic insulators *Phys. Rev. B* **2** 1324-1338

Brodimas G, Janussis A and Mignani R 1992 Bose realizations of a non-canonical Heisenberg algebra *J. Phys. A* **25** L329-L334

Bruschi M, Levi D and Ragnisco O 1981 Evolution equations associated with the discrete analog of the matrix Schrödinger problem solvable by the inverse spectral transform *J. Math. Phys.* **22** 2463-2471

Büttiker M 1985 Small normal-metal loop coupled to an electron reservoir *Phys. Rev.* B **32** 1846-1849

Byers N and Yang C N 1961 Theoretical considerations concerning quantized magnetic flux in superconducting cylinders *Phys. Rev. Lett.* **7** 46-49

Caldirola P 1976 On the introduction of a fundamental interval of time in quantum mechanics *Lett. Nuovo Cimento* **16** 151-155

Carazza B and Kragh H 1995 Heisenberg's lattice world: The 1930 theory sketch *Am. J. Phys.* **63** 595-605

Carow-Watamura U and Watamura S 1994 The q-deformed Schrödinger equation of the harmonic oscillator on the quantum Euclidean space, *Int. J. Mod. Phys. A* **9** 3989-4008

Carvalho Dias F, Pimentel I R and Henkel M 2006 Persistent current and Drude weight in mesoscopic rings *Phys. Rev.* B **73** 075109 pp. 1-7

Celeghini E, S. De Martino, S. De Siena, Rasetti, M and Vitiello G 1995 Quantum groups, coherent states squeezing and lattice quantum mechanics *Ann. Phys.* **241** 50-67

Chabaud E, Gallinar J-P and Mata J 1986 The quantum harmonic oscillator on a lattice *J. Phys.*A **19**, L385-L390

Chaichian M, Sheikh-Jabhari M M and Tureanu A 2001 Hydrogen atom spectrum and the Lamb shift in noncommutative QED *Phys. Rev. Lett.* **86** 2716-2719.

Chair N and Dalabeeh M A 2005 The noncomutative quadratic Stark effect for the H-atom *J. Phys. A* **38** 1553-1556.

Chakraborti JT and Pictilainen P 1994 Electron-electron interaction and the persistent current in a quantum ring *Phys. Rev.* B **50** 8460-8468.

Chakraborti J 1995 Dirac Hamiltonian on a lattice, *Int. J. Mod. Phys.* A **14** 2087-2095

Chapard B and Drozi M 1998 *Cellular Automata Modelling of Physical Systems*(Australia: Cambridge Univ. Press)

Chari V and Pressley A 1994 *A guide to quantum groups* (Cambridge: Cambridge Univ. Press)

Chen B, Dai X and Han R 2002 Dynamic localization in mesoscopic metallic rings *Phys. Lett.* A **302** 325-329

Chen T W, Huang C M and Guo Y G 2006 Conserved spin and orbital angular momentum Hall current in a two-dimensional electron system with Rashba and Dresselhaus spin-obit coupling *Phys. Rev.* B **73** 2353091 pp. 1-13

Chen Y, Xiong S J and Evangelou S N 1997 Quantum oscillations in mesoscopic rings with many chains *Phys. Rev.* B **54** 4778-4785

Cheung H F, Gefen Y, Riedel E K and Shih W H 1988 Persistent currents in small one-dimensional metal rings *Phys. Rev.* B **37** 6050-6062

Chung W S 1993 q-deformed Baker-Campbell-Hausdorff formula *Mod. Phys. Lett. A* **8** 2569-2572

Citrin D S 2004 Magnetic Bloch Oscillations in Nanowire Superlattice Rings, *Phys. Rev. Lett.* **12** 196803 pp. 1-4

Coleman P 1984 New approach to the mixed valence problem *Phys. Rev.* B **29**, 3035-3044

Comtet A 1987 On the Landau levels on a hyperbolic plane *Ann. Phys.* **173** 185-209

Connes A 1994 *Noncommutative geometry* (New York: Academic Press)

Curado E M, Rego-Monteiro M A and Nazareno H N 2001 Heisenberg-type structures of one-dimensional quantum Hamiltonians *Phys. Rev.* A **64** 012105 pp.1-5

Czererwinski M and Brown E 1991 Generalized gauge invariance and the integer quantum Hall effect *Proc. Roy. Soc. London* A **432** 43-54

Darboux G 1889 *Lecons sur la theorie generale des surfaces et les applications geometriques du calcul infinitesimal, II* (Paris: Gautier-Villars)

Datta S 1995 *Electronic transport in mesoscopic systems* (Cambridge: Cambridge University Press)

Davison S G and Steslicka M 1992 *Basic Theory of Surface States* (Oxford: Clarendon Press)

De Moura F A B F and Lyra M L 1998 Delocalization in the 1D Anderson-model with long-range correlated disorder *Phys. Rev. Lett.* **81** 3735-3738

Debergh N, Ndimubandi J and Bosche van den B 2002 A general approach of quasi-exactly solvable Schrödinger-equations *Ann. Phys.* **298** 361-381

De Oliveira C R and Prado R A 2005 Dynamical delocalization for the 1D Bernoulli discrete operator *J. Phys. A***38** L115-L119.

De Souza Doutra A 1993 Conditionally exactly soluble class of quantum potentials *Phys. Rev. A* **47** R2435-R2437

Deych L I, Erementchouk M V and Lisyansky A A 2003 Scaling in the one-dimensional Anderson localization problem in the region of fluctuation states *Phys. Rev. Lett.* **90** 126601 pp. 1-3

Dignam M M and de Sterke C M 2002 Conditions for dynamic localization in generalized ac electric fields *Phys. Rev. Lett.***88** 046806 pp. 1-4.

Doebner H D and Goldin G A 1992 On a general nonlinear Schrödinger-equation admitting diffusion coefficients *Phys. Lett. A* **162** 397-401

Domachuk P, Martijn de Sterke C, Wan J and Dignem M M 2002 Dynamic localization in continuous ac electric fields *Phys. Rev. B* **66** 165313 pp. 1-10

Dominguez-Adame F, Malyshev V A, de Moura F A B F and Lyra M L 2003 Bloch like oscillations in a one dimensional lattice with long range correlated disorder *Phys. Rev. Lett.* **91**, 197402 pp. 1-4

Dunlap D H and Kenkre V M 1986 Dynamic localization of a charged particle moving under the influence of an electric field, *Phys. Rev.* B **34** 3625-3633

Dyakin V V and Petrukhnovskii D I 1986 On discreteness of the spectrum of some operator sheaves associated with a periodic Schrödinger equation *Theor. Math. Phys.* **74** 66-72

Eckmann J P and Ruelle D 1985 Ergodic theory of chaos and strange attractors *Rev. Mod. Phys.* **57** 617-656

Enolskii V I and Eilbeck J C 1995 On the two-gap locus for the elliptic Calogero-Moser model *J. Phys. A* **28** 1069-1088

Evangelou S N and Pichard J L 2000 Critical quantum chaos and the one-dimensional Harper model *Phys. Rev. Lett.* **84** 1643-1646

Faddeev L D, Reshetikin N Yu and Takhtajan L A 1990 Quantization of Lie groups and Lie algebras *Leningrad Math. J.* **1** 193-225

Faddeev L D and Kashaev R M 1995 Generalized Bethe ansatz equations for Hofstadter problem *Commun. Math. Phys.* **169** 181-191

Fano U 1961 Effects of configuration interaction on intensities and phase shifts *Phys. Rev.* **124** 1866-1878

Feigenbaum J and Freund P G O 1996 A q deformation of the Coulomb-problem *J. Math. Phys.* **37** 1602-1616

Ferrari V, Chiappe G, Anda E and Davidovich M A 1999 Kondo resonance Effect on persistent Currents through a Quantum Dot in a Mesoscopic Ring *Phys.*

Rev. Lett. **82** 5088-5091

Finkel F, Gónzales-López A and Rodriguez M A 2000 A new algebraization of the Lamé equation *J. Phys.* A **33** 1519-1542

Finkelstein D 1997 *Quantum relativity. A synthesis of ideas of Einstein and Heisenberg* (Berlin: Springer).

Floreanini R and Vinet L 1995a Lie symmetries of finite difference equations *J. Math. Phys.* **36** 7024-7042

Floreanini R and Vinet L 1995b Quantum symmetries of q-difference equations *J. Math. Phys.* **36** 3134-3156

Flores J C 2005 Quantum LC circuits with charge discretness: Normal and anomalous spectrum *Europhys. Lett.* **69** 116-120.

Flügge S 1971 *Practical Quantum Mechanics I*, (Berlin: Berlin Springer)

Fradkin E 1991 *Field Theories of Condensate Matter Systems*(see chapter 9)(Redwood City: Addison Wesley)

Freeman M, Mateev M D and Mir-Kasimov R M 1969 On a relativistic quasipotential equation with local interaction *Nucl. Phys. B* **12** 197-215

Fukuyama H, Bari R A and Fogedby H C 1973 Tightly bound electrons in a uniform electric field *Phys. Rev.* B **8** 5579-5586

Fye R M, Martins M J, Scalapino D J, Wagner J and Hanke W 1991 Drude weight optical conductance and flux properties of one-dimensional Hubbard rings *Phys. Rev.* B **44** 6909-6915

Gallinar J-P 1984 The Coulomb potential problem on the Bethe lattice *Phys. Lett.* A **103** 72-74

Gallinar J-P and Mattis D C 1985 Motion of "hoping" particles in a constant force field *J. Phys.* A **18** 2583-2590

Ganguly A 2002 Associated Lamé and various other new classes of elliptic potentials from $sl(2, R)$ and related orthogonal polynomials *J. Math. Phys.* **43** 1980-1999

Gasper G and Rahman M 1990 *Basic Hypergeometric Series*(Cambridge: Cambridge,Univ. Press)

Gaudin M 1983 *La fonction d'onde de Bethe*(Paris: Masson)

Gebhard F 1997 *The Mott metal-insulator transitions - models and methods*(Heidelberg: Springer Verlag)

Geisel T, Ketzmerick R and Petschel G 1991 New class of level statistics in quantum systems with unbounded diffusion *Phys. Rev. Lett.* **66** 1651-1654

Gerhardts R R, Wein D and Wulf U 1991 Magnetoresistance oscillations in a grid potential: Indication of a Hofstadter-type energy spectrum *Phys. Rev.* B **43** 5192-5195

Gómez I, Dominguez-Adame F and Orellana P A 2004 Fano-like resonances in three quantum-dot Aharonov-Bohm rings *J. Phys. Cond. Matter* **16** 1613-1621

Gradshteyn I S and Ryzhik I M 1965 *Tables of Integrals, Series and Products* (New York: Academic Press)

Grandy W T Jr and Goulart Rosa S Jr 1981 Applications of Mellin transforms to the statistical mechanics of ideal quantum gases *Am. J. Phys.* **49** 570-578

Gredeskul S A, Zusman M, Avishai Y and Azbel M Ya 1997 Spectral properties and localization of an electron in a two-dimensional system with point scatterers in a magnetic field *Phys. Rep.* **288** 223-257

Grempel D R, Fishman S and Prange R E 1982 Localization in an incommensurate potential: an exactly solvable model *Phys. Rev. Lett.***49** 833-836

Guedes I 2001 Solution of the Schrödinger equation for the time dependent linear potential *Phys. Rev. A* **63** 034102 pp. 1-3

Gusynin V P and Sharapov S G 2005 Unconventional integer quantum Hall effect in graphene *Phys. Rev. Lett.* **95** 146801 pp. 1-4

Halburd R G 2005 Diophantine integrability *J. Phys. A* **38** L263-L269

Halperin B I 1984 Statistics of the quasiparticles and the hierarchy of fractional quantized Hall states *Phys. Rev. Lett.* **52** 1583-1586

Hanein Y, Shahar D, Yoon J, Li C C, Tsui D C and Shtrikman H 1998 Observation of the metal-insulator transition in two-dimensional n-type GaAs *Phys. Rev.* B **58** R13334-R13340

Harper P G 1955 Single band motion of conduction electrons in a uniform magnetic field *Proc. Phys. Soc. London* A **68** 874-878

Hasegawa Y, Hatsugai Y, Kohmoto M and Montambaux G 1990 Stabilization of flux states on two-dimensional lattices *Phys. Rev.* B **41** 9174-9181

Hatsugai Y and Kohmoto M 1990 Energy spectrum and the quantum Hall effect on a square lattice with next-nearest-neighbor *Phys. Rev.* B **42** 8282-8294

Hatsugai Y 1993 Edge states in the integer quantum Hall effect and the Riemann surface of the Bloch function *Phys. Rev.* B **48** 11851-11862

Hatsugai Y, Kohmoto M and Wu Y S 1994 Explicit solutions of the Bethe ansatz equations for Bloch electrons in a magnetic field *Phys. Rev. Lett.* **73** 1134-1137

Hayashi A, Hashimoto T, Horibe M and Yamamoto H 1997 New formulation of the Dirac equation on a Minkowsky lattice *Phys. Rev.* D **55** 2987-2993

Hebecker A and Weich W 1992 Free particle in q-deformed configuration space, *Lett. Math. Phys.* **26** 245-248

Heisenberg W 1938a Über die in der Theorie der Elementarteilchen auftretende universelle Länge *Ann. Phys. Lpz.* **32** 20-33

Heisenberg W 1938b Die Grenzen der Anwendbarkeit der bisherigen Quantentheorie *Z. Phys.* **110** 251-266

Hellund E J and Tanaka K 1954 Quantized space-time *Phys. Rev.* **94** 192-195

Hewson A C 1993 *The Kondo Problem to Heavy Fermions* (Cambridge: Cambridge University Press)

Hilke M 2003 Noninteracting electrons and the metal insulator transition in two dimensions *Phys. Rev. Lett.* **91** 226403 pp. 1-4

Hiramoto H and Kohmoto M 1989 Scaling analysis of quasiperiodic systems: generalized Harper model *Phys. Rev.* B **40** 8225-8234

Hirsch J E 1999 Spin Hall Effect *Phys. Rev. Lett.* **83** 1834-1837

Hoffmann T 2000 On the equivalence of the discrete nonlinear Schrödinger equation and the discrete isotropic Heisenberg magnet *Phys. Lett.* A **265** 62-67

Hofstadter D R 1976 Energy levels and wave functions of Bloch electrons in rational and irrational magnetic fields *Phys. Rev.* B **14** 2239-2249

Holthaus M 1992 Collapse of minibands in far-infrared irradiated superlattices

Phys. Rev. Lett. **69** 351-354

Holthaus M and Hone D 1993 Quantum wells and superlattices in strong time-dependent fields *Phys. Rev.* B **47** 6499-6508

Honda T and Tokihiro T 1997 Spontaneous lattice deformation in the Hofstadter system *Phys. Rev.* B **55** 10261-10269

Hong S P and Salk S H S 1999 Harper's equation for two-dimensional systems of antiferromagnetically correlated electrons *Phys. Rev.* B **60** 9550-9554

Hubbard J 1979 Electronic structure of one-dimensional alloys *Phys. Rev.* B **19** 1828-1839

Huckenstein B 1995 Scaling theory of the integer quantum Hall effect *Rev. Mod. Phys.* **67** 357-396

Hufnagel L, Ketzmerick R, Kottis T and Geisel T 2001 Superballistic spreading of wave packets *Phys. Rev.* E **64** 012301 pp. 1-3.

Infeld L and Hull T 1951 The factorization method *Rev. Mod. Phys.* **23** 21-68

Ishikawa K and Maeda N 1997 Flux state in von Neumann lattices and the fractional Hall-effect *Prog. Theor. Phys.* **97** 507-525

Izrailev F M, Ruffo S and Tessieri L 1998 Classical representation of one-dimensional Anderson model *J. Phys.* A **31** 5263-5270

Jackson F H 1909 Generalization of the differential operative symbol with an extended form of Boole's equation *Messenger. Math.* **38** 57-61

Jackson F H 1910 On q-definite integrals *Quart. J. Pure Appl. Math.* **41** 193-203

Jaksch D and Zoller P 2003 Creation of effective magnetic fields in optical lattices: the Hofstadter butterfly for cold neutral atoms *New Journal of Physics* **5** 56.1-56.11

Janussis A 1984 Quantum equations of motion in the finite difference approach *Lett. Nuovo Cimento* **40** 250-256

Janussis A and Brodimas G 2000 Connection between q-algebras and noncanonical Heisenberg algebras *Mod. Phys. Lett.* A **15** 1385-1390

Jatkar D P Kumar C N and Khare A 1989 A quasi-exactly solvable problem without SL(2) symmetry *Phys. Lett.* A **142** 200-202

Jellal A 2001 Orbital magnetism of a two dimensional noncommutative confined system *J. Phys. A* **34** 10159-10177

Jivulescu M A and Papp E 2006 On the dynamic localization conditions for dc-ac electric fields proceeding beyond the nearest neighbor description, *J. Phys. Condens. Matter* **8** 6853-6857

Kadyshevsky V G 1961 On the theory of quantization of space-time *Zh. Eksp. Teor. Fiz.* **41** 1885-1894

Kadyshevsky V G, Mir-Kasimov R M and Nagiyev Sh M 1968 Quasi-potential approach and the expansion in relativistic spherical functions *Nuovo Cimento A* **55** 233-257

Kagramanov E D, Mir-Kasimov R M and Nagiyev Sh M 1990 The covariant linear oscillator and generalized realization of dynamical $SU(1,1)$ symmetry algebra *J. Math. Phys.* **31** 1733-1738

Kang K and Shin S C 2000 Mesoscopic Kondo Effect in Aharonov-Bohm ring *Phys. Rev. Lett.* **85** 5619-5622

Karlo T, Jacob H and Tripathy K C 1994 q calculus and the discrete inverse scattering *Mod. Phys. Lett.* A **10** 3021-3032

Kehagias A A and Zoupanos G 1994 Finiteness due to cellular structure of R^NI. Quantum mechanics *Z. Phys.* C **62** 121-126

Kempf A, Mangano G and Mann R B 1995 Hilbert space representation of the minimal length uncertainty relation *Phys. Rev.* D **52** 1108-1118

Kenkre V M, Kühne R and Reineker P 1981 Connection to the velocity autocorrelation function to the mean square displacement and to the memory function of generalized master equations *Z. Phys. Cond. Matter* B **41** 177-180

Ketzmerick R, Kruse K, Springsguth D and Geisel T 2000 Bloch Electrons in a Magnetic Field: Why Does Chaos Send Electrons the Hard Way? *Phys. Rev. Lett.* **84** 2929-2932

Keyser U F, Fühner C, Borch S, Haug J R, Bichler M, Abstreiter G and Wegscheider W 2003 Kondo effect in a few electron quantum ring *Phys. Rev. Lett.* **90** 196601 pp. 1-4

Khare A and Sukhatme U 2006 Periodic potentials and PT symmetry *J.Phys.* A **39** 10133-10142

Kishigi K and Hasegawa Y 2002 De Haas-van Alphen effect in two-dimensional systems *Phys. Rev. B* **65** 205405 pp. 1-9

Klappauf B G, Oskay W H, Steck D A and Raizen M G 1998 Observation of noise and dissipation effects on dynamical localization *Phys. Rev. Lett.* **81** 1203-1206

Klitzing von K, Dorda G and Pepper M 1980 New method for high-accuracy determination of the fine-structure constant based on quantized Hall resistance *Phys. Rev. Lett.* **45** 494-497

Klitzing von K 1986 The quantized Hall effect *Rev. Mod. Phys.* **58** 519-531

Knox R S 1963 *Excitons* (New York: Academic Press)

Kobayashi K, Aikawa H, Katsumoto S and Iye Y 2004 Mesoscopic Fano effect through a quantum dot in an Aharonov-Bohm ring *Physica* E **22** 468-473

Kohmoto M, Kadanoff L P and Tang C 1983 Localization problem in one dimension: mapping and escape *Phys. Rev. Lett.* **50** 1870-1872

Kohmoto M 1989 Zero modes and the quantized Hall conductance of the two-dimensional lattice in a magnetic field *Phys. Rev.* B **39** 11943-11949

Kohmoto M and Hatsugai Y 1990 Peierls stabilization of magnetic-flux states of two-dimensional lattice electrons *Phys. Rev.* B **41** 9527-9529

Korepin V E, Izergin A G and Bogoliubov N N 1992 *Quantum Inverse Scattering Method and Correlation Functions* (Cambridge: Cambridge Univ. Press)

Kostelecky V A, Nieto M M and Truax D R 1985 Supersymmetry and the relationship between the Coulomb and oscillator problems in arbitrary dimensions *Phys. Rev.* D **32** 2627-2633

Kostelecky V A and Russell N 1996 Radial Coulomb and oscillator systems in arbitrary dimensions *J. Math. Phys.* **37** 2166-2181

Kramer B and MacKinnon A 1993 Localization: theory and experiment *Rep. Prog. Phys.* **56** 1469-1564

Krasovsky I V 1999 Bethe ansatz for the Harper equation: solution for a small commensurability parameter *Phys. Rev.* **59** 322-328

Kravchenko S V, Kravchenko G V, Furneaux J E, Pudalov V M and D'Iorio M 1994 Possible metal-insulator transitions at B=0 in two dimensions *Phys. Rev.* B **50** 8039-8042

Krawtchouk M 1929 Sur une géné ralisation des polinomes d'Hermite *Comp. Rend. Acad. Sci. Paris,* **189** 620-622

Kroon L, Lennholm F and Ricklund R 2002 Localization-delocalization in aperiodic systems, *Phys. Rev.* B **66** 094204 pp. 1-9

Kubo R 1966 The fluctuation-dissipation theorem *Rep. Progr. Phys.* **29** 255-284

Kulish P P and Damaskinsky E V 1990 On the q oscillator and the quantum algebra $su_q(1,1)$ *J. Phys.* A **23** L415-L419

Kvitsinsky A A 1992 Exact solution of the Coulomb problem on a lattice *J. Phys.* A **25** 65-72

Kvitsinsky A A 1994 An exactly solvable class of discrete Schrödinger equations *J. Phys.* A **27** 215-218

Lai K, Pan W, Tsui D C, Lyon S A Mühlberger M and Schäffler F 2005 Two-dimesional metal-insulator transition and in-plane magnetoresistence in a hihg-mobility strained Si quantum well *Phys. Rev.* B **72** R 81313 pp. 1-4

Landauer R and Büttiker M 1985 Resistance of small metallic loops *Phys. Rev. Lett.* **54** 2049-2052

Last Y and Wilkinson M 1992 A sum rule for the dispersion relations of the rational Harper equation *J. Phys.* A **25** 6123-6133

Lee T D 1983 Can time be a discrete dynamical variable? *Phys. Lett. B* **122** 217-220

Lévi B and Georgeot B 2004 Quantum computation of a complex system: The kicked Harper model *Phys. Rev.* E **70** 056218 pp. 1-19

Levi D and Rodriguez M A 1999 Lie symmetries for integrable evolution equations on the lattice *J. Phys.* A **32** 8303-8316

Li B and Kwok D Y 2003 Discrete Boltzmann equation for microfluids *Phys. Rev. Lett.* **90** 124502 pp. 1-4

Li H and Kusnezov D 2000a Dynamical symmetry approach to periodic Hamiltonians *J. Math. Phys.* **41** 2706-2722

Li H, Kusnezov D and Iachello F 2000b Group theoretical properties and band structure of the Lamé-Hamiltonian *J. Phys. A* **33** 6413-6429

Li Y Q and Sheng Z M 1992 A deformation of quantum mechanics *J. Phys. A* **25** 6779-6788. Erratum: *J. Phys. A* **28** 3565

Lindquist B and Riklund R 1998 Deterministic aperiodic one-dimensional systems with all states extended, one of which is periodic *J. Phys. Soc. Jpn.* **67** 1672-1676

Lipan O 2000 Bandwidth statistics from the eigenvalue moments for the Harper-Hofstadter problem *J. Phys.* A **33** 6875-6888

Liu Y and Hui P M 1998 Electronic transport properties of tight-binding multiring systems *Phys. Rev.* B **57** 12994-13001.

Lorente M 1989 On some integrable one-dimensional quantum mechanical systems *Phys. Lett.* B **232** 345-350.

Lorente M and Kramer P 1999 Representations of the discrete inhomogeneous Lorentz group and Dirac wave equation on the lattice *J. Phys.* A **32** 2481-

2497

Lorente M 2001a Raising and lowering operators, factorization and differential/difference operators of hypergeometric type *J. Phys.* A **34** 569-588

Lorente M 2001b Continuous vs. discrete models for the quantum harmonic oscillator and the hydrogen atom *Phys. Lett.* A **285** 119-126

Louisell W H 1973 *Quantum Statistical Properties of Radiation* (New York: John-Willey)

Luck J M 1989 Cantor spectra and scaling of gap widths in deterministic aperiodic systems *Phys. Rev.* B **39** 5834-5849

MacAnnally D S 1995 q-exponential and q gamma functions. I. q-exponential functions *J. Math. Phys.* **36** 546-573

Macfarlané A J 1989 On q-analogues of the quantum harmonic oscillator and the quantum group $SU_q(2)$ *J. Phys. A* **22** 4581-4588

Madureira J R, Schulz P A and Maialle M Z 2004 Dynamic localization in finite quantum dot superlattices: A pure ac field effect *Phys. Rev.* B **70** 033309 pp. 1-4

Magnusdottir I and Gudmundsson V 1999 Influence of the shape of quantum dots on their far-infraraed absorbtion *Phys. Rev.* B **60** 16591-16617.

Mailly D, Chapellier C and Benoit A 1993 experimental observation of persistent currents in GaAs-AlGaAs single loop *Phys. Rev. Lett.* **70** 2020-2023

Majid S 1990 Quasitriangular Hopf algebras and Yang-Baxter equations *Int. J. Mod. Phys.* A **5** 1-91

Manin Yu I 1988 Quantum groups and noncommutative geometry, *Les Publ. de Recherche Math.* Université de Montreal.

March A 1937 Die Frage nach der Existenz einer kleinsten Wellenlänge *Z. Phys.* **108** 128-136.

Micu C and Papp E 2003 Applying the virial and Hellmann-Feynman theorems to the generalized q-symmetrized Harper-equation *Phys. Lett. A* **314** 10-14

Micu M 1999 q-deformed Schrödinger-equation *J. Phys. A* **32** 7765-7777

Migliore R and Messina A 2004 Freezing the dynamics of a rf SQUID qubit via its strong coupling to a quantized microwave field *J. Opt. B: Quantum Semiclass. Opt.* **6** S136-S141

Miller Jr W 1969 Lie theory and difference equations I *J. Math. Anal. Appl.* **28** 383-399

Miller Jr W 1972 Lie theory and difference equations II *J. Math. Anal. Appl.* **29** 406-422

Miller Jr W 1970 Lie theory and q-difference equations *SIAM J. Math. Anal.* **1** 171-188

Minot C 2004 Quantum model of electronic transport in superlattice minibands *Phys. Rev.* B **70** R 161309 pp. 1-4

Mir-Kasimov R M 1991 $SU_q(1,1)$ and the relativistic oscillator *J. Phys. A* **24** 4283-4302

Moore F L, Robinson J C Bhancha C, Williams P E and Raizen M G 1994 Observation of dynamical localization in atomic momentum transfer: a new testing ground for chaos *Phys. Rev. Lett.* **73** 2974-2977

Morita Y and Hatsugai Y 2001 Duality in the Azbel-Hofstadter problem and the two-dimensional d-wave superconductivity with a magnetic field *Phys. Rev. Lett.* **86** 151-154

Muñoz E, Barticevic Z and Pacheco M 2005 Electronic spectrum of a two-dimensional quantum dot array in the presence of electric and magnetic fields in the Hall configuration *Phys. Rev.* B **71** 165301 pp. 1-9

Nathanson B Entin-Wohlmann O and Muhlschlegel B 1998 Comment on "Peierls gap in a mesoscopic ring threaded by a magnetic flux" *Phys. Rev. Lett.* **80** 3416

Nazareno H N and de Brito P E 2001 Carriers in a two-dimensional lattice under magnetic and electric fields *Phys. Rev.* B **64** 045112 pp. 1-6

Negro J and Nieto L M 1996 Symmetries of the wave equation in a uniform lattice *J. Phys.* A **29** 1107-1114

Nelson C A and Gartley M G 1994 On the zeros of the q-analogue exponential function *J. Phys. A* **27** 3857-3881

Németh Z A and Picard J L 2005 Persistent currents in two dimensions: New regimes induced by the interplay between correlations and disorder *Eur. Phys. J.* B **45** 111-128

Nenciu G 1991 Dynamics of band electrons in electric and magnetic fields: rigorous justification of the effective Hamiltonians *Rev. Mod. Phys.* **63** 91-127

Nielsen M A and Chuang I L 2000 *Quantum computation and quantum information* (Cambridge: Cambridge University Press)

Nieto M M 1979 Hydrogen atom and relativistic pi-mesic atom in N-space dimensions *Am. J. Phys.* **47** 1067-1072

Nikiforov A F, Suslov S K and Uvarov V B 1991 *Classical Orthogonal Polynomials of a discrete variable*(Berlin: Springer)

Niu Q 1989 Effect of an electric field on a split Bloch band *Phys. Rev.* B **40** 3625-3637

Nouicer K 2006 An exact solution of the one-dimensional Dirac oscillator in the presence of minimal lengths *J. Phys.* A **39** 5125-5134

Nussenzweig H M 1972*Causality and Dispersion Relations*(New York: Academic Press)

Obermair G M and Wannier G H 1976 Bloch electrons in magnetic fields. Rationality, irrationality, degeneracy *Phys. Stat. Sol.* B **76** 217-222

Ocampo S 1996 $SO_q(4)$ quantum mechanics *Z. Phys. C.* **70** 525-530

Orellana P A , Lara G A and Anda E V 2002 Kondo effect and bistability in a double quantum dot *Phys. Rev.* B **65**, 155317 1-5

Orellana P A, Dominguez-Adame F, Gomez I and Ladrón de Guevara M L 2003a Transport through a quantum wire with a side quantum-dot array *Phys. Rev.* B **67** 085321 pp. 1-5

Orellana P A, Ladrón de Guevara M L, Pacheco M and Latgé A 2003b Conductance and persistent current of a quantum ring coupled to a quantum wire under external fields *Phys. Rev.* B **68** 195321 pp. 1-7

Osadchy D and Avron J E 2001 Hofstadter butterfly as a quantum phase diagram *J. Math. Phys.* **43** 5665-5671

Ostlund S, Pandit R, Raud D, Schellnhuber H J and Siggia E D 1983a One-dimensional Schrödinger equation with an almost periodic potential *Phys.*

Rev. Lett. **50** 1873-1876

Ostlund S, Rand D, Sethna J and Siggia E 1983b Universal properties of the transition from quasi-periodicity to chaos in dissipative systems *Physica* D **8** 303-342

Ostlund S and Pandit R 1984 Renormalization group analysis of the discrete quasiperiodic Schrödinger equation *Phys. Rev.* B **29** 1394-1414

Ott E 1994 *Chaos in Dynamical Systems* (Cambridge:Cambridge Univ. Press)

Pan Hui-yun and Zhao Z S 2001 Operator realizations of the q-deformed Heisenberg algebra *Phys. Lett.* A **282** 251-256

Papadakis S J and Shayegan M 1998 Apparent metallic behavior at B=0 of a two-dimensional system in AlAs *Phys. Rev.* B **57** R15068-R15071

Papp E 1997 *Derivation of q-analogs for the radial Schrödinger equation in N space dimensions* (New York: Nova Science)

Papp E Micu C and Szakacs Zs 2002a Revisiting density of state calculations for the Harper equation *Int. J. Mod. Phys.* B **16** 3481-3489

Papp E and Micu C 2002b Deriving exact energy solutions to the symmetrized q-difference Harper equation *Phys. Rev.* E **65** 046234 pp. 1-8

Papp E 2003 Influence of the anisotropy parameter Δ on the spectrum of the generalized q-symmetrized Harper equation *J. Phys.* A **36** 2077-2086

Papp E and Micu C 2005 *Quantum systems on discrete spaces-Theoretical descriptions and applications to nanoscaled structures* (Cluj: Risoprint)

Papp E 2006 Time discretization approach to dynamic localization conditions *Int.J.Mod.Phys.*B **20**, 2237-2254

Papp E, Aur L, Micu C and Racolta D 2006 Period doubling effects in the oscillations of persisten currents in discretized Aharonov-Bohm rings *Physica E* (to be published)

Peierls R 1933 Der Diamagnetismus der Leitungselektronen *Z. Phys.* **80** 763-791

Peter D, Cyrot M, Mayou D and Khanna S N 1989 Kinetic energy of electrons on a two-dimensional lattice with commensurate flux *Phys. Rev.* B **40** 9382-9384

Petschel G and Geisel T 1993 Bloch electrons in magnetic fields: Classical chaos and Hofstadter butterfly *Phys. Rev. Lett.* **71** 239-242

Pfannkuche D and Gerhardts R R 1992 Theory of magnetotransport in two-dimensional electron systems subjected to weak two-dimensional superlattice potentials *Phys. Rev.* B **46** 12606-12626

Piéchon F 1996 Anomalous diffusion properties of wave packets in quasiperiodic chains *Phys. Rev. Lett.* **76** 4372-4375

Planck M 1913 *Vorlesungen über die Theorie der Wärmestrahlung* (Leipzig: J. A. Barth) see the last chapter

Poincarè H 1913 *Dernières pensés* (Paris: Flammarion)

Pokrowski G I 1928 Zur Frage nach der Struktur der Zeit *Z. Phys.* **51** 737-739

Pronin K A, Bandrauk A D and Ovchinnikov A A 1994 Harmonic generation by a one-dimensional conductor: Exact results *Phys. Rev.* B **50** R 3473-3476

Quieroz De S L A 2002 Failure of single parameter scaling of wavefunctions in Anderson localization *Phys. Rev.* B **66** 195113 pp. 1-5

Raimondi R and Schwab P 2005 Spin Hall effect in disordered two-dimensional electron system *Phys. Rev.* B **71** 033311 pp. 1-4

Ramaswamy R 2002 Symmetry-breaking in local Lyapunov exponents *Eur. Phys. J.* B **29** 339-343

Rangarajan G, Habib S and Ryne R D 1998 Lyapunov exponents without rescaling and reorthogonalization *Phys. Rev. Lett.* **80** 3747-3750

Rauh A 1975 On the broadening of Landau levels in crystals *Phys. Stat. Sol.* B **69** K9-K13

Ringot J, Szriftgiser P Garreau J C and Delande D 2000 Experimental evidence of dynamical localization and delocalization in a quasiperiodic driven system *Phys. Rev. Lett.* **85** 2741-2744

Romerio M V 1971 Almost periodic functions and the theory of disordered systems *J. Math. Phys.* **12** 552-562

Ruuska V and Törmä P 2004 Quantum transport of non-interacting Fermi gas in an optical lattice combined with harmonic trapping *New Journal of Physics* **6** 59.1-59.11

Savrasov S Y and Kotliar G 2003 Linear response calculations of lattice dynamics in strongly correlated systems *Phys. Rev. Lett.* **90** 056401 pp. 1-4

Seiberg N and Witten N 1999 String theory and noncommutative geometry *J. High Energy Phys.* **32** 09, 1-93

Shafiekhani A 1994 $U_q(sl(n))$ difference operator realization *Mod. Phys. Lett.* A **9** 3273-3283

Shifman M A 1989a New findings in quantum mechanics (partial algebraization of the spectrum) *Int. J. Mod. Phys.* A **4** 2897-2952

Shifman M A and Turbiner A V 1989b Quantal problems with partial algebraization of the spectrum *Commun. Math. Phys.* **126** 347-365

Shifman M A and Turbiner A V 1999 Energy-reflection symmetry of Lie-algebraic problems: where the quasiclassical and weak-coupling expansions meet *Phys. Rev.* A **59** 1791-1798

Sil S, Karmakar S N and Moitra R K 1993 Extended electronic states of disordered 1-d lattices: an example *Phys. Lett.*A **183** 344-347

Simon B 1982 Almost periodic Schrödinger operators *Adv. Appl. Math.* **3** 463-479

Simon B 1985 Almost periodic Schrödinger operators IV. The Maryland model *Ann. Phys.* **159** 157-183

Simonin J, Proetto C R, Barticevic Z and Fuster B 2004 Single particle electronic spectra of quantum rings: A comparative study *Phys. Rev.* B **70** 205305 pp. 1-8.

Smirnov Yu and Turbiner A V 1995 Lie algebraic discretization of differential equations *Mod. Phys. Lett.* A **10** 1795-1802. Erratum: *Mod. Phys. Lett.* A **10** pp. 3139

Snyder N S 1947 Quantized space-time *Phys. Rev.* **71** 38-41

Sokoloff J B 1981 Quasiclassical theory of quantum particles in two incommensurate periodic potentials *Phys. Rev.* B **23** 2039-2041

Sokoloff J B 1985 Unusual band structure, wave functions and electrical conductance in crystals with incommensurate periodic potentials *Phys. Rep.* **126** 189-244

Song X C and Liao L 1992 The quantum Schrödinger-equation and the q-deformation of the hydrogen atom, *J. Phys. A* **25** 623-634

Springsguth D, Ketzmerick R and Geisel T 1997 Hall conductance of Bloch electrons in a magnetic field *Phys. Rev.* B **56** 2036-2043

Spiridonov V, Vinet L and Zhedanov A 1993 Difference Schrödinger operators with linear and exponential discrete spectra *Lett. Math. Phys.* **29** 63-73

Streda P 1982 Theory of quantized Hall conductivity in two dimensions *J. Phys.* C **15** L717-L721

Stormer H L, Schlesinger Z, Chang A, Tsui D C, Gossard A C and Wiegmannn W 1983 Energy Structure and Quantized Hall Effect of Two-Dimensional Holes *Phys. Rev. Lett.* **51** 126-129

Suqing D, Wang Z G, Wu B Y and Zhao X G 2003 Dynamical localization of semiconductor superlattices under a dc-bychromatic electric field *Phys. Lett.* A **320** 63-69

Tan W C and Inkson J C 1999 Magnetization, persistent currents and their relation in quantum rings and dots *Phys. Rev.* B **60** 5626-5635.

Tan Y 1995 Total bandwidth of the Harper equation: connection to renormalization analysis *J. Phys.* A **28** 4163-4173

Thouless D J 1972 A relation between the density of states and range of localization for one dimensional random system *J. Phys.* C **5** 77-81

Thouless D J, Kohmoto M, Nightingale M P and M den Nijs 1982 Quantized Hall conductance in a two-dimensional periodic potential *Phys. Rev. Lett.* **49** 405-408

Thouless D J 1983a Bandwidth for a quasiperiodic tight-binding model *Phys. Rev.* B **28** 4272-4276

Thouless D J 1983b Quantization of particle transport *Phys. Rev.*B **27** 6083-6087

Torres M and Kunold A 2004 Dynamical localization of Bloch electrons in magnetic and electric fields *Phys. Lett.* A **323** 290-297

Tsui D C, Stormer H L and Gossard A C 1982 Two-dimensional magnetotransport in the extreme quantum limit, *Phys. Rev. Lett.* **48** 1559-1562

Turbiner A V and Ushveridze A G 1987 Spectral singularities and quasi-exactly solvable quantal problem *Phys. Lett.* A **126** 181-183

Turbiner A V 1992 Lie algebras and polynomials in one variable *J. Phys.* A **25** 1087-1094

Twarock R 1999 A q-Schrödinger equation based on a Hopf q-deformation of the Witt algebra *J. Phys. A* **34** 4971-4981

Ubriaco M R 1992 Time evolution in quantum mechanics on the quantum line *Phys. Lett. A* **163** 1-4

Ubriaco M R 1993 Quantum deformations of quantum mechanics, *Mod. Phys. Lett. A* **8** 89-96

Ushveridze A G 1994 *Quasi Exactly Solvable Models in Quantum Mechanics* (Bristol: Institute of Physics)

Vialtzew A N 1963 *Discrete space and time*(Moskwa: Nauka) (in russian)

Vidal J, Mosseri R and Douçot B 1998 Aharonov-Bohm cages in two-dimensional structures *Phys. Rev. Lett.* **81** 5888-5891

Vilenkin N Ya and Klimyk A V 1992 *Representations of Lie groups and special functions. Vol. 3: Classical and quantum groups and special functions* (Dordrecht: Kluwer Academic)

Voisin P, Bleuse J, Bouche C, Gaillard S, Alibert C and Regreby A 1988 Observation of Wannier-Stark quantization in a semiconductor supperlattice *Phys. Rev. Lett.* **61** 1639-1642

Wacker A, Jauho A P, Zeuner S and Allen S. J. 1998 Sequential tunneling in doped superlattices: Fingerprints of impurity bands and photon assisted tunneling *Phys. Rev.* B **56** 13268-13278.

Wacker A 2002 Semiconductor superlattices: a model system for nonlinear transport *Phys. Rep.* **357** 1-111.

Wan J, de Sterke C M Dignam M M 2004 Dynamic localization and quasi-Bloch oscillations in general periodic ac-dc electric fields *Phys. Rev.* B **70** 125311 pp. 1-4

Wang X F and Vasilopoulos P 2003 Magnetotransport in a two-dimensional electron gas in the presence of spin-orbit interaction *Phys. Rev.* B **67** 085313 pp. 1-7

Wang Y Y, Pannetier B and Rammal R 1987 Quasiclassical approximations for almost-Mathieu equations *J. Physique* **48** 2067-2079

Wannier G H 1960 Wave functions and effective Hamiltonian for Bloch electrons in an electric field, *Phys. Rev.* **117** 432-439

Wannier G H 1975 Invariance properties of a proposed Hamiltonian for Bloch electrons in a magnetic field *Phys. Stat. Sol.* B **70** 727-735

Wannier G H 1978 A result not depending on rationality for Bloch electrons in a magnetic field *Phys. Stat. Sol.* B **88** 757-765

Wannier G H, Obermair G and Ray R 1979 Magnetoelectronic density of states for a model crystal *Phys. Stat. Sol.* B **93** 337-342

Watanabe S, Strogatz S H, van der Zant H S J and Orlando T O 1995 Whirling modes and parametric instabilities in the discrete Sine-Gordon equation: experimental tests in Josephson rings *Phys. Rev. Lett.* **74** 379-382

Wess J 1997 Q-deformed phase space and its lattice structure *Int. J. Mod. Phys.* A **12** 4997-5005

Wheeler J A 1957 On the nature of quantum geometrodynamics *Ann. Phys.* (N.Y.) **2** 604-614

Wiegmann P B and Zabrodin A 1994a Bethe-ansatz for Bloch electron in a magnetic field *Phys. Rev. Lett.* **72** 1890-1893

Wiegmann P B and Zabrodin A 1994b Quantum group and magnetic translations Bethe ansatz for the Azbel-Hofstadter problem *Nucl. Phys.* B **422** 495-514

Wiegmann P B and Zabrodin A V 1995 Algebraization of difference eigenvalue equations related to $U_q(sl(2))$ *Nucl. Phys.* B **451** 699-724

Wilkinson M 1986 Von Neumann lattices of Wannier functions for Bloch electrons in a magnetic field *Proc. Roy. Soc. London* A **403** 135-166

Wilkinson M and Austin E J 1994 Spectral dimension and dynamics for Harper's equation *Phys. Rev.* B **50** 1420-1429

Wineland D J, Bergquist J C, Itano W M, Bollinger J J and Manney C H 1987 Atomic in Coulomb clusters in a trap *Phys. Rev. Lett.* **26** 2935-2938

Wolf C 1989 Modification of the mass momentum energy relation due to nonlocal discrete time quantum effects *Nuovo Cimento* B **103** 649-655

Wolfram S 1986 *Theory and applications of cellular automata* (Singapore: World

Scientific)

Wuerker R F, Shelton H and Langmuir R V 1959 Electromagnetic containment of charged particles *J. Appl. Phys.* **30** 342-349

Xiong Y T and Liang X T 2004 Fano resonance and persistent current of a quantum ring *Phys. Lett.* A **330** 307-312

Xiong Y T and Liang X T 2005 The phase of Fano resonance in Aharonov-Bohm interferometer with a quantum dot embedded *Physica* B **355** 216-221

Yamani H A and Fishman L 1975 *J* matrix method: extensions to arbitrary angular momentum and to Coulomb scattering *J. Math. Phys.* **16** 410-420

Yang C N 1947 On quantized space-time *Phys. Rev.* **72** p. 874

Yang Q G and Xu B W 1993 Energy spectrum of a q -analogue of the hydrogen atom *J. Phys. A* **26** L365-L368

Zak J 1964 Group theoretical consideration of Landau level broadening in crystals *Phys. Rev.* **136** A776-A780

Zhang A Z, Zhang P, Suqing D, Zhao X G and Liang J Q 2002 Dynamics of a superlattice with an impurity in a dc-ac electric field *Phys. Lett.* A **292** 275-280

Zhang X W and Xia J B 2006 Rashba spin-orbit copling in InSb nanowires under tranverse electric field *Phys. Rev.* B **74** 075304 pp. 1-8

Zhao X G, Jahnke R and Niu Q 1995 Dynamic fractional Stark ladders in dc-ac fields *Phys. Lett.* A **202** 297-304

Zhao X G, Georgakis G A and Niu Q 1997 Photon assisted transport in superlattices beyond the nearest neighbor apppoximation *Phys. Rev.* B **56** 3976-3981

Zhedanov A S 1993 On the realization of the Weyl commutation relation HR=qRH *Phys. Lett.* A **176** 300-302

Zhu M J, Zhao X G and Niu Q 1999 Manipulations of band electrons with a rectangular wave electric field *J. Phys. Condens. Matter* **11** 4521-4538

Ziman J M 1964 *Principles of the theory of solids* (Cambridge: University Press)

Zumino B 1991 Deformation of the quantum mechanical phase space with bosonic or fermionic coordinates *Mod. Phys. Lett.* A **6** 1225-1235

Index

1D discretized ring, 170, 172
1D lattice, 221
1D ring, 34, 35, 40
2D lattice, 133
2D periodic potential, 143, 149

Aharonov-Bohm boundary condition, 210
Aharonov-Bohm effect, 195
Aharonov-Bohm magnetic flux, 34, 38
Aharonov-Bohm potential, 30, 32
Airy-functions, 62
Anderson model, 107
anisotropy parameter, 152, 175
annihilation operators, 45
area quantization condition, 122
average ring current, 204

bandwidth, 22, 160
Bessel-function, 63
Bessel-functions, 62
Bethe-ansatz, 79, 237
Bloch states, 208
Bloch-electrons, 1, 175
Bloch-period, 172
Bloch-theorem, 17, 19, 23–26, 30
Boltzmann equation, 131
boundary conditions, 102, 232
Brillouin phases, 139, 152, 153
Brillouin zone, 17, 23, 59, 63, 111, 125, 136, 140, 209

canonical commutation relations, 221
Casimir-eigenvalue, 187
Casimir-operator, 84, 186
chain of quantum dots, 207
Charlier polynomials, 3
chemical potential, 162
classical LC circuit, 215
classical limit, 86, 87, 90, 91
classical orthogonal polynomials of a discrete variable, 14
coherence length, 210
commensurability parameter, 134
Compton-wavelength, 22
concavity condition, 119, 220
conductance, 205, 210
Coulomb-gauge, 175
Coulomb-problem, 66, 89–91
creation operators, 45

deformation parameter, 1, 69
density of states, 103–105, 157, 159–161, 163, 165
density operator, 48
difference equation, 223
Diophantine equation, 166
discrete analog of the harmonic oscillator, 87
discrete analog of the radial Coulomb system, 89, 91
discrete derivative, 41, 216
discrete derivatives, 2, 13
discrete equation, 3, 4, 7, 67, 69, 75,

76, 87, 89, 91, 92, 99, 103, 153, 176, 212, 219
discrete Fourier transformation, 142
discrete integral, 137
discrete Schroedinger equation, 77, 97, 99
discrete Schroedinger-equation, 57, 61, 65
discrete space, 1, 9–11, 13, 14, 17, 26, 28, 87
discrete time, 41, 48, 55
discrete variable, 1, 7
discretized Aharonov-Bohm ring, 196, 199
double quantum dot system, 220, 222
duality transformations, 143
dynamic localization, 107–109, 113, 218, 219
dynamic localization condition, 114, 115, 119, 120, 122–124, 126, 220

effective Hamiltonian, 218
elliptic integral, 159
energy of the incident electron, 204
energy reflection symmetry, 183
energy-band, 17, 19, 21, 139
energy-dispersion law, 66, 152, 176
energy-polynomials, 157
exactly solvable systems, 82, 83
exponential potential, 69, 70

Fermi Dirac distribution, 166
Fermi-energy, 165
finite difference Liouville-von Neumann equation, 48
fixed boundary conditions, 102
Floquet-factorization, 127
flux quantum, 215
Fock states, 46, 47

gauge parameters, 175
generators of rotations, 10
Green-function, 165

Hall conductance, 140, 165, 183
Hall conductivity, 165, 166
Hamiltonian of hypergeometric type, 83
harmonic oscillator, 24, 84, 87, 216, 218
Harper-equation, 151–153, 182
Heisenberg algebra, 80
Heisenberg representation, 42
Heisenberg-Weyl commutation relation, 79, 95
Hellmann-Feynman theorem, 224
Hermitian space-time generators, 10
Hofstadter butterfly, 157
hopping amplitude, 199, 212
hopping Hamiltonian, 99, 133
hopping integral, 99, 133, 135
hopping interaction, 222
hopping parameter, 199, 200, 207, 222
Hulthen potentials, 69
hypergeometric function, 236

integrated density of states, 103, 104, 160, 165
inter-band couplings, 146
inter-dot tunneling coupling, 222
iterative mapping, 104

Jackson q-integral, 6
Jackson derivative, 3, 5

Kondo regime, 220
Krawtchouk polynomials, 3, 14, 84–87, 232, 233
Kubo's formula, 140
Kummer-function, 64

Lame-equation, 27
Landau-gauge, 133, 152, 175, 192
Landauer formula, 210
Leibniz-rule, 2, 3
level broadening, 210
Lie algebraic approach, 82
linear potential, 62
linear second-order discrete equations, 208
localization length, 106, 151, 154

localized orthonormalized
  Wannier-states, 100
lowering operator, 233
Lyapunov exponents, 103, 104, 154,
  160

magnetic Bloch oscillations, 172
magnetic Brillouin zone, 192
magnetic flux operator, 217
magnetic flux quantum, 133
magnetic length, 31
magnetic phase factor, 133
magnetic translation operator, 189,
  190, 192
magnetic translations, 152, 188, 192
magnetization, 162, 163, 183
Maryland-model, 69, 70
Mathieu-equation, 59, 61
Mathieu-type equation, 218
Mathieu-wavefunctions, 60
mean square displacement, 109,
  111–113, 118–120, 122, 219, 220
Meixner-polynomials, 14, 89, 90, 232
Mellin-transform, 164
multichain nanorings, 210

nanoconductor, 108
nanoring, 152, 167
nanoscale systems, 28
normalization, 138, 180, 182, 232
normalized Krawtchouk-functions, 85
normalized Meixner-function, 94
normalized Meixner-functions, 89, 91
normalized wavefunction, 101

orthogonality condition, 97, 142
oscillator wave function, 143

parabolic cylinder function, 47
periodic AB boundary condition, 35,
  173
periodic behavior, 18, 26
periodic boundary condition, 201
periodic potentials, 17, 19, 99, 100
persistent current, 31, 33, 35, 40, 173,
  205, 219
persistent currents, 195

q-binomial theorem, 4
q-deformed algebras, 79
q-deformed
  Baker-Campbell-Hausdorff formula,
  52
q-deformed equations, 5
q-difference equation, 237
q-Gamma function, 5
q-hypergeometric function, 9
q-integral, 5
q-normalization, 180
q-shifted factorial, 5
q-symmetrized Harper-equation, *175*,
  188
quantum dot, 222
quantum dot potential, 28
quantum L ring circuit, 115, 219
quantum LC circuit, 195, 215, 218
quantum ring potential, 31
quantum wire, 207, 210
quasi-energy, 126, 128–130
quasi-monomials, 80
quasiperiodic potentials, 25, 99, 100

raising operator, 233
recurrence relation, 9, 44, 45, 70, 178,
  179, 233
reflection amplitude, 209
relaxation time approximation, 131
rescaled amplitude, 219
rescaled Meixner-function, 93, 94
rescaled parameter, 223
ring energy, 204
Rodriguez formula, 231

second order discrete equation, 218
second-order discrete equation, 240
secular equation, 139, 182
shifted factorial, 5, 8
sinusoidal Fourier-series, 196
SL(2)-symmetry, 82, 187, 188
SL(N)-group, 79
slave boson description, 221

SLq(2)-symmetry, 83, 95, 183, 185, 186
SLq(N)-symmetry, 79
SO(3)-symmetry, 84
SOq(N)-symmetry, 5
spectral density, 107
SUq(2)-symmetry, 191
symmetrized derivative, 4
symmetrized momentum operator, 217

thermodynamic Hall conductance, 160
thermodynamic potential, 162
Thouless-formula, 105
tight binding, 79
tight binding equation, 146
tight binding Hamiltonian, 222
time dependent current, 132
total current, 196
total persistent current, 35–37, 169, 174
trace formula, 154
transfer matrix, 153, 182
transmission amplitude, 209
transmission coefficient, 210
transmission probability, 205, 215
tridiagonal matrix, 138
tuning parameter, 201
tunneling Hamiltonian, 201
tunneling interaction, 207, 211
two dimensional electron gas, 151, 161

unions over Q's, 157
unitary operators, 41

Wannier states, 207, 223
Wannier-Stark ladders, 126
wavefunction ansatz, 209
weight function, 14, 15, 233

zero temperature magnetization, 33